A New Earth Rising

CHARMIAN AMAREA KUMARA REDWOOD

For permission, serialization, condensation, adaptions, or for our catalog of other publications, write to Ozark Mountain Publishing, Inc., P.O. Box 754, Huntsville, AR 72740, ATTN: Permissions Department.

Library of Congress Cataloging -in-Publication Data

Redwood, Charmian, 1951

A New Earth Rising by Charmian Redwood

Charmian shares information about the steps we need to take in order to prepare ourselves and our loved ones to move smoothly into the new world of Peace.

1. New Earth 2. Hypnosis 3. Earth Changes 4. Metaphysical

I. Redwood, Charmian, 1951   II. New Earth III. Metaphysical IV. Hypnosis V. Title

Library of Congress Catalog Card Number: 2015940869

ISBN : 9781940265155

Cover Art and Layout: www.noir33.com
Book set in: Lucida Fax, Lucida Handwriting
Book Design: Tab Pillar

Published by:

OZARK
MOUNTAIN
PUBLISHING

PO Box 754

Huntsville, AR 72740

WWW.OZARKMT.COM

Printed in the United States of America

To my beloved twins, light of my life, and also to all of those who are holding the light as we make this transition from darkness into light.

Table of Contents

Introduction:
A New Earth Is Being Born   i

Chapter 1:
Hypnotherapy and How the Information Arrived for This Book 1

Chapter 2:
Coming Home to Yourself   5

Chapter 3:
The Galactic Alignment of December 21, 2012   16

Chapter 4:
Lifetimes of Preparation for Our Return to Oneness   22

Chapter 5:
Earth's Final Clearing from Energies of Control & Resistance   31

Chapter 6:
The New Children Who Came to Help Birth the New Earth   43

Chapter 7:
Beings from Many Dimensions Now Embodied on Earth   54

Chapter 8:
A New Earth Rising   63

Chapter 9:
Help Comes from the Stars to Make This Dimensional Shift   76

Chapter 10:
Emergence, Inner Earth, and Beings from Telos Come to Assist 86

Chapter 11:
The New Way of Living from Joy   94

Chapter 12:
Releasing Old Karmic Relationships   100

Chapter 13:
Awakening to a New World   106

Chapter 14:
How the Shift of Earth's Axis Will Affect Those Here   114

Chapter 15:
Return of the Christ   122

Chapter 16:
Earth Changes Are Preparing the Way   127

Chapter 17:
Peace Shall Reign   131

Chapter 18:
Eye Hath Not Seen   133
Conclusion   137
Contributors   139
About the Author   143

Introduction

A New Earth Is Being Born

This is a time of great transformation on the Earth. We are at the end of the cycle of darkness and separation that began with The Fall from Oneness into density in the time of Lemuria. The Mayans said that on December 21, 2012, time as we know it would end. This did not mean that the Earth would end but that the way that we have been living, in suffering, separation, and pain would end. There was much fear on the planet as some people were predicting the end of the world so I was inspired to use my skills as a hypnotherapist to progress some people forward through the Dimensional Shift and into the New Earth to see what was actually going to be happening on and after that date.

We have now passed through that gateway of December 21, 2012, and Earth has indeed shifted her vibration completely out of the third dimension and into the fifth. This change of frequency is causing vast and rapid change for us humans. Since that time we have been through a great cleansing. Karma from the third dimension ended on December 21 as the Earth moved out of the dimension that has linear time. We are no longer bound by time and space so any residue from our third-dimensional experience must be released now from the body. None of the structures or belief systems from the third dimension has any validity anymore. We are sailing new uncharted seas with our own God Presence at the helm.

Now we need to release all of the old emotions and negativity that would hold us in separation. If we hold onto them, they toxify both our own bodies and those whom we project them onto. A good way to do this is to use the ancient Hawaiian ceremony of Ho'opono pono, where we ask forgiveness of everybody for anything we may have done that has harmed them in thought, word, or deed. We then forgive them and ourselves for anything they have done to us.

It is essential that we release all old feelings of anger, resentment, self-denial, lack of self-love, and grief both toward ourselves and others. All of our structures both internally and in the systems that support our outer lives (such as politics and banking) are now being realigned with the principles of Oneness. Old systems must dissolve so that they can be recreated in harmony with the Law of One. So we will see much change over the next few years, but we must remember that everything is working for the return of the planet to Love.

I was at a retreat in Mount Shasta at the time of the shift. I awoke spontaneously at 3:33 on the morning of December 21, 2012, at the exact time of the predicted shift. I felt a soft blue light come down over the whole planet as gently as a snowflake. It was the energy of the Cosmic Christ coming down, not with fireworks and explosions as I was expecting, but gently. I felt this soft blue light envelop my whole body very easily and completely. I went back to sleep and when I woke up again later I was wondering which of the many events in Shasta for the shift I should attend. That was when I felt the energy and realized what the shift had been. I knew without a doubt that it didn't matter if I went to all or none of the events because I had shifted into my Being-ness. This meant that wherever I went from this time forward I was in my Presence and would take it with me wherever I was. I had embodied my Christed I Am Presence and from now on I only need to "Be" this Presence wherever I am, nothing more to do. With such a liberation, I didn't go to any of the events; instead, I had a lovely massage and sang Christmas carols in the local cafe.

Before the shift happened I had just published my first book, *Coming Home to Lemuria*, which was about the Golden Age of peace and harmony with which we began our journey on the Earth plane. I was getting a lot of e-mails from all over the world from people who were in fear about the coming changes upon the Earth. They were wondering if they and their loved ones were going to survive the transition from the Old World to the New. In response to this anxiety I guided many people forward in time using hypnotherapy to look at life on the Earth after the Dimensional Shift, and this book is the result. These are actual people. I have used their names with their permission if they agreed and have used an abbreviation for the names of those who wished to remain anonymous.

Each chapter is the story of one person. It is obvious from this information that we have prepared for many lifetimes both on the Earth and on other star systems to be able to complete this transformation and to return to the Golden Age of Peace on the Earth. Memories of ancient Lemuria are coming forward now because we have already lived on the Earth in Oneness and harmony, and we will do so again. These memories are helping us to remember who we are as Divine Beings in human form.

We have been living a third-dimensional existence on this planet for thousands of years where we have focused on the material, physical plane, believing that only things we can actually see with our eyes or experience with our senses are real. There are many planes above the physical one that have been invisible to us as we lived in density. Now that the frequency of the Earth is rising and the centers within the brain that allow us to perceive the more subtle planes of the fourth and fifth dimensions are being activated, the higher planes of these dimension are becoming accessible to us. The fifth-dimensional plane is above and beyond separation or duality. There is no good and bad; there is only love, light, and Oneness where we absolutely know that we are part of God, as is all of Creation. Quantum physics is now beginning to rediscover what mystics and tribal people have known for eons, that all life is made of the same substance, it all has consciousness and is interrelated.

The Earth has been preparing for decades to make an unprecedented energy shift from the third dimension to the fifth dimension where everything and everyone on her surface will be operating from their God consciousness. This process is called the Ascension, the Resurrection, or the Dimensional Shift. Living in the old way in the third dimension we have been living from greed, misuse of power, separation, and "the few" controlling the lives of "the many" as well as most of the resources. In the New Way this will no longer be possible, as only systems that are heart based will function on the Earth in her new frequency.

Many of the old structures, which have supported the misuse of power, are crumbling. The banking system, financial institutions, and big corporations with political influence are and will continue to destabilize. New systems of equal distribution of wealth and resources will take their place. Everything that has been happening with the economy, the financial systems, and the many changes of government throughout the world are part of this process, which is absolutely necessary. We simply cannot continue to live upon the Earth under the present system of corruption, disregard for human rights, deprivation of basic necessities, and pollution of our Mother Earth. So the old order is crumbling, being burned in the fires of transformation, and out of the ashes a New Earth is rising, a return to an age of gold where peace shall reign in the hearts of men and women.

Part of this process involves a cleansing of the Earth by all four elements (fire, water, earth, and air) as Mother Earth prepares her body to receive the higher frequencies of pure light and pure love. Many people saw in the hypnotherapy sessions that there could be some dramatic geological events such as earthquakes, active volcanoes, hurricanes, tornadoes, and tsunamis but only in places where people had become very disconnected from the values of community and sharing or in places where there was a lot of negative energy. They also saw that the intensity of these events could be modified if we start now to work with Mother Earth by showing her love and appreciation for the gifts that she pours upon us. This is a book about hope and a New World being born, not about destruction and the end of the world.

I was shown many years ago that there were two spirals of energy happening on the Earth right now. One is an ascending spiral that is taking matter back to spirit and light, and the other is a descending spiral that is the energy of separation, control, and limitation. The descending one has nowhere to go and will eventually disappear into itself. We are all very powerful God Beings, and each one of us needs to choose where we will put our energy, support, and power. The descending spiral appears to be doing a great job of destroying both itself and the Old Order. I have always known that I came to build the New Earth rather than to participate in the destruction of the old one. So I choose to put my energy into the ascending spiral that is taking us home.

In the beginning Mother Earth asked to become a fifth-dimensional planet, and she sent out a request to the Light-workers and to the children of the Sun to come and walk upon her body to impregnate her with the light. Many of those walking upon her body today are Lightworkers who responded to her call and came down from the fifth dimension and above to assist in this transformation process. We first came down to Earth as the civilization known as Lemuria in our bodies of light. We lived in our light bodies, not in physical dense bodies, and we created everything we needed from the ethers by focusing our intention through the heart and literally manifesting our world from the energy of the Source. We came from the Light. We were the Light. During that time we knew only love and service, and everything we did was for the good of the whole and of all kingdoms. (See my book *Coming Home to Lemuria*.)

Then came The Fall where we became disconnected from Source. We forgot that we were Divine Beings here on the Earth and entered many thousands of years of density, suffering, and pain. Before The Fall we lived in Oneness, connection, and harmony. We did not know anything except love and harmony. We did not have disconnection so there was no possibility of conflict. We had no concept of conflict or any frame of reference for such a concept. Then we chose, as a way to advance the evolution of our souls, to experience being disconnected from the Source. We would go down into density where we had never been before, we would forget who we were, we would go to sleep, and we would go through this time of suffering and conflict. Then at the end we would return to the Oneness and bring the Earth with us.

That is where we are right now. We are at the jumping off point where everything and everybody is being lifted back up into the light. The Earth has moved energetically into the fifth dimension so the time of separation is officially ended and the Earth is returning to the Garden of Eden. The plan always was that we would come from the Light; we would descend into duality and separation; we would experience ourselves as separate entities; and then we would return to the union with our own God Presence that had always remained in the Light. We were the ones who made the plan, and the timing was written from the beginning. During the transition from the Age of Pisces, which was an age of sacrifice to the Age of Aquarius, which is an age of God-Realization, we would move out of density and separation back to light and Oneness on the Earth.

On December 21, 2012, the Earth aligned exactly with the center of our galaxy and with the great central sun, and this alignment sent a huge surge of energy across the planet. This is the event that we had been waiting for. This surge of energy activated the light bodies of those who are ready to receive it, and this prepared the way for the first wave of Ascension. There will be a series of waves of Ascension for a certain time upon the Earth. When these waves are complete, those souls who have chosen not to make the Ascension will be moved to other planets where they can continue their experience at that level. There is no judgment in this. It is just a matter of what each soul chooses for its own evolutionary process.

Through the time of density we as humans have only brought 30–40 percent of our soul's energy into our body. After the Ascension we will have 100 percent of our soul in the body as we had in ancient Lemuria. Then we will be able to operate from our own God consciousness; we will be fully Self-Realized Beings. We will be able to communicate telepathically with the animals, the angels, our star brothers and sisters, and all Beings on all of the dimensional planes. We will have the ability to teleport ourselves wherever we want to go simply by seeing ourselves as already there.

After the Dimensional Shift we will eventually be operating from our Higher Self or light body but will still be in physical form. That is what the shift is. We will be fully conscious God-Realized Beings using 100 percent of our brains. The parts of the brain that have been dormant for thousands of years are the parts that connect us to our Higher Self. Those parts are being reactivated now, and the strands of our DNA that were shut down for the descent into darkness are being reconnected. We have only been using a 2-stranded DNA that just allows us to function at a survival level. After the shift we will once more have access to our full 22-strand DNA and will be using what are now thought of as supernatural (but which are actually quite normal) powers in the fifth dimension. That is why so many of the children's video games are about overcoming challenges and collecting powers; they are helping them to remember who they are.

After our individual and collective Ascension we will be living on the Earth as Christed Beings. Every single person who makes the Ascension will be a fully Christed Being. This is the meaning of the Second Coming. We are moving into Christhood or Resurrection as Yeshua came to demonstrate. Then we will rebuild the Earth and clean up all the pollution of the air and water in an instant with our newly acquired powers. We will be operating in our full consciousness as Divine humans. We will be fully connected to our Higher Selves that are always connected to the Source. Our Higher Self knows the plan. It knows why we are here. It is helping us to fulfill our soul's purpose and is bringing us all of the necessary life experiences that we need in order to be prepared to download our light body into our physical body and to make the connection with the Higher Self.

Concerning the lower self, the personality, or ego self, when we make the energy shift we will have full connection with our Higher Self. Then the ego self will drop away. We will be able to see the bigger picture in everything and not be so reactive to situations from our old ego patterns and beliefs.

It has been a wild ride here on planet Earth. All of the solar flares that are coming from the sun are like having a fire hose with fire instead of water blasting energy at the Earth. Our bodies are electromagnetic; we are really just electrical impulses, and we are getting short-circuited. More power is coming through our energy systems than we are used to. This is how we are being awakened and activated. Gradually over the last twenty years we have had increasingly high-frequency energy coming down to the Earth, and now we have gone through the gateway of December 21, 2012, which was a major activation. This was the trigger for the first wave of the Ascension for millions of people on Earth, and there are a series of supercharged gateways that will assist in preparing for this activation.

The spring equinox 2012 was an opportunity to come into balance and harmony, and the solar eclipse in May 2012 opened a gateway for connection to the God Self. Many areas of our planet are already in the fifth dimension. During the equinoxes of the last three years leading up to the Ascension, the threshold to higher dimensionality has opened much wider.

11.11.11 was a huge activation of the crystal grids that anchored the blueprints for Ascension into the Earth grids. The ceremonies, prayers, and focused groups that we as Lightworkers have been doing since then have fed directly into that grid. The 11.11.11 activation opened up the gateway for Ascension. All of the energy that we have been bringing in since then has added more complexity to that grid, which strengthens the Ascension frequency. We need to focus on feeding love, harmony, and peace into the grid, which will help to offset the places on the Earth that are feeding on conflict and hatred such as the Middle East and Africa.

The global chakra system that is part of the grid was activated by the 11.11.11 event. The vortex points of Mount Shasta, Sedona, Stonehenge, Glastonbury, Giza, Machu Picchu, Ayers Rock, and Lake Titicaca are now all holding the frequencies for the Ascension of the Earth into her new body of light. It is most important for Lightworkers to feed their energy into the grids at these critical vortex points. We, the Lightworkers, are so powerful, we are like crystalline acupuncture needles, and everything that we are feeling is going directly into the crystalline Earth grid feeding the blueprint for Ascension, wholeness, and awakening. Many of us found ourselves traveling to these powerful energy portals to receive the new frequencies and also to upgrade them from our own vibration.

The Venus transit in May/June 2012 also triggered more shifts as we head into our date with destiny in December. Venus connects our whole solar system with the vibration of Love, Beauty, and the Divine Feminine. These are the energies that are rising on the Earth at this powerful time and we are being drawn toward the expression Love and Beauty in our Lives. Anything that is not vibrationally aligned with these rising frequencies is going to come under stress at this time.

The solar eclipse of May 20 marked the "point of no return" on our journey toward the December solstice 2012. The more ability we have to release emotional attachments, thoughts, and ideologies that hold us back from experiencing full conscious awareness of our connection to the Source, the Oneness of all Life, the smoother the transition will be for us. That eclipse brought a golden opportunity to bring all aspects of our lives into balance and right relation.

It helped us to stand fully in the power of who we truly are and what we truly believe.

This high-frequency energy is having a very powerful effect on our physical and emotional bodies. People are having all kinds of "Ascension symptoms" from headaches, sleeplessness, and high temperatures to anxiety attacks, depression, and fear. These may also be experienced as mood swings, dizziness, insomnia, lethargy, apathy, and fatigue. Each new upgrade of energy requires physical, emotional, and mental adjustment to the higher frequencies downloaded.

Every trauma from our previous lives in the long cycle since The Fall is being triggered now so that it can be released. You may find in your personal life that anything that is residual in your energy body from the 25,000 years of descent into darkness is being brought to your attention in dramatic ways. You may feel disoriented or just overwhelmed, but you are not alone. It is important to connect with likeminded people and share what is happening for you. Having bodywork or acupuncture and taking sea salt baths will help to clear all the residue that is being brought to the surface in your energy field. You may think you are going crazy, but this too shall pass.

We are literally growing a new body that will be much better suited to holding our light bodies that are vibrating at a vastly higher frequency than before. For this reason we are changing the basic cellular structure of our physical bodies from a carbon to a silicone base; the molecular structure is changing, which means that every system in our body is being recalibrated. This is why so many people are having issues with blood, bones, nerves, muscles, eyesight, and sudden surges of body heat. The Earth is growing a new body hence the extreme geological and climactic events. We are growing a new body hence all the physical, emotional, and mental issues. Our social systems are also growing a new body hence the collapse of our financial and economic foundation. We simply can't put a new suit of clothes onto a worn-out vehicle that wasn't built for the purpose. It is all interconnected—what we do to the Earth, we do to ourselves, and what we do to each other affects the Earth.

Not everyone is making this shift. Many people will be leaving, and you will hear about people transitioning through illness, accidents, or war. The gateway is open, but not all of those who are on the planet at this time will choose to make this Dimensional Shift. Those people will eventually leave the planet. After a certain period only those who are operating in the fifth dimension will still be on the Earth because she will be in her fifth-dimensional body, and third-dimensional life will no longer be supported here. There will be a transition time where some people are bringing in the new way of peace, love, and harmony, and others are still trying to hold onto the old ways of abusive power, greed, and division. It will be increasingly difficult to maintain the old structures that kept people separate and disempowered. Life for those who try to hold onto the old ways will become evermore chaotic and unsupportable. The awakened ones will be helping those who are still making the choice to embrace the new way.

There are millions of Light Beings here to help us, including Star Beings, Angels, and Ascended Masters. There never has been a time like this on any planet when we have gone from such a dark place into full light in such an incredibly short time. Heaven is coming to Earth, and we are bringing it here. For every individual on the Earth, it is time to make a choice, to say, "Yes, I am here to serve the Earth and this transformation." We are the midwives, and Mother Earth is birthing her light body. The way that she births, whether it is easy or difficult, is very much up to us. If we are consciously working with her and offering ourselves as vessels to be of service, then her transition will be smooth. If we remain in ignorance and continue to cause her suffering by polluting her waters, air, and soil, then she will create events that will restore the balance and cause much destruction.

I was in Nevada at the time of hurricane "Sandy" and was due to fly to Washington Dulles airport on the Wednesday that was predicted to be at the height of the hurricane. I did a journey to ask what the storm was about. I saw a huge lion storming up the east coast with his teeth and claws bared. He passed by the coast of Florida, South Carolina, and Virginia and then very deliberately turned toward land in New York and Boston. He was very angry at the damage that humans were causing Mother Earth. He wanted the people in the places where decisions are made that harm the land and the water to experience how it would be if Mother Earth decided to use us in the same way we are abusing her. It wasn't punishment; it was to make us aware that she is alive and we need to understand that and respect her. I stood to the side of him and asked if there was anything we could do to help. He looked at me in amazement that a human would even think to offer help. He said, "Yes, you can pray and you can tell the people Mother is hurting." So I prayed and I also send out e-mails asking others to pray. Many tribal elders were also doing prayer ceremonies for Earth Mother. Within hours the hurricane dropped down to a storm and caused much less damage than was predicted. I visited with the lion the next day; he was sitting on the ocean purring, his claws were in, and there was a dove of peace sitting on his back. Mother Earth loves us. She wants us all to live in harmony, but she can no longer tolerate the suffering that our ignorance is causing to all the kingdoms.

I have been noticing for several years that whenever I hear about a potentially catastrophic event coming, such as a hurricane or a tsunami, I immediately go into a place of prayer and ask of Mother Earth, "Would you mind softening the effect of this event? I know you need to get our attention because it is part of your cleansing, but please would you do it gently?" The effect is immediate, and within hours the potential storm has calmed.

There was a hurricane heading toward Hilo on the Big Island of Hawaii about five years ago when I was there; an e-mail went out asking people to work with the hurricane. I immediately went into prayer and visualized the hurricane diminishing. Within the next five hours it went down to a tropical storm and blew over the island with no damage at all. There was also a tsunami that came to Hawaii and as we worked with its energy it diminished, and by the time it arrived it was reported as being about two inches high. That is what we can do. That is the power that we have. It is up to us to make the choice of what we do with that power. Each of us is a Living Master, a Christed Being in the making, and what we do with our power will decide the manner in which Mother Earth makes this transition.

I was living in Hawaii where the waves can be very strong. I kept getting knocked over when I was coming out of the ocean, either on my paddleboard or swimming. After trying many methods I learned that the best way to get out safely was to talk to Mother Ocean and say, "Please let me out gently," and she always did. While I am out in the ocean,

I say prayers and sing to her, and she helps me to get out gently. I mention this because the changes on the Earth are a gradual process, not a one-time event, and there are likely to be many geological and weather events as we move forward into the new paradigm. So never think that we are victims. We are now co-creators and have the power to change anything. This is one of the things that happened at the time of the Dimensional Shift on December 21. We moved out of "victim" into co-creation where we simply focus our intention and can move mountains. Our brains moved out of functioning in alpha frequency and reactivity into theta frequency and creatorship.

The dimensional shift of December 21, 2012, officially moved the Earth out of the third dimension and into the fifth. The time of duality and separation has ended, and we are now in the process of shedding all of the accumulated negativity from our whole traverse through density. The lunar and solar eclipses of May 2013 as well as the powerful solar flares at that time were huge clearing opportunities to cleanse our emotional bodies of debris from our time of suffering and to activate more of our solar light body. When enough people have lifted their frequency into their Christed I Am Presence, then humanity will take a quantum jump into the New Earth where we will be living from our Divine Essence.

Chapter 1

Hypnotherapy and How the Information Arrived for This Book

Incarnation is like taking on a character in a play. Your soul has a home, and that home is in the Golden Light, the Source. You choose what you want to do and what experiences you want to have in order for your soul to grow.

My purpose for being on the Earth is to help people to remember that what we are experiencing here on Earth is just a play. It's a story: we made it up, it's not real, and it's just an illusion. What is real is your soul or your Higher Self. Your soul is always knocking on your door trying to get your attention, and usually that happens through an illness, or in my case I actually died, went to heaven, and came back. Some people have accidents that are supposed to wake them up to their soul's calling.

In my near-death experience I left my body behind and went home as a soul to the golden light. I experienced myself as a Being of Pure light. When I came back into my body I asked for a way to help people to experience themselves as souls while they are still alive. A lot of people have that experience on their deathbed. They realize that they could have had a very different life if they had known at the beginning what they know at the end—that you don't die when the body dies. The means that was given to me in answer to that request was to use hypnotherapy to guide people to connect with the Oneness.

I use transpersonal hypnosis to take people into a deep trance state where they meet themselves as Divine Essence. They look at the plan for their life. They look at why they chose to come into this body at this time on the Earth. Everybody that is in your life now is someone you have a soul contract with. You are playing out a script that you wrote together with your soul family before you came into the body. We all have a soul family, and we meet up again at the

1

end of this lifetime when we leave our bodies. We review the life that we have just completed, and we look at whether or not we fulfilled our soul's mission. If we feel that we did not fulfill our soul's purpose, we look at how we want to go back and do it better. If we took on karma from things that we didn't do very well or where we caused harmed, we decide how we can repay the karmic debt.

Different people in the same soul group will come back as mother and son one time, husband and wife another time, or adversaries another time. Someone has to agree to be our challenger. Our challengers are our teachers; they are the ones who help us grow. We keep coming around in all of these different stories and roles, and it is all to help the soul to evolve. Every incarnation brings our souls up to another level. If it is a very difficult lifetime, you have the opportunity to grow a lot.

In my hypnosis sessions I help people find out why they are here at this time, what is the reason for events that are happening in their lives, and what might be coming in the future. It really helps people come to terms with situations that make them angry, sad, or frustrated. When you understand the bigger picture, you don't get so involved with the small stuff. It is working in the psyche at the level of cause not effect. Hypnosis helps the whole person to come into integration, harmony, and balance. It helps people to feel better and gives them hope when they feel hopeless; from understanding come serenity, acceptance, and hope. So that is my mission and purpose, to give people a reason to carry on.

When you are in a deep trance, the brain is in a theta wave state, the deep-sleep frequency. The conscious mind goes to sleep, and it is as if you are talking to your own Divine Essence. Your Divine Essence knows everything, the plan for the planet, all of your lifetimes, and all of your contracts. You just ask your own Inner Being whatever you want to know.

In the sessions receiving information for this book, *A New Earth Rising*, I guided people to connect with their own Higher Self and then asked that Self if it would guide the rest of the session and give us information about the coming Dimensional Shift.

Using this hypnotic technique I guided many people through the Dimensional Shift that began in December 21,

2012, and continues in the following years. We have passed this date, which was the activation of the new cycle and the end of separation on the Earth. The information that different people brought through was remarkably similar. We looked at the New Earth and what it will be like. It is like in Revelations: "I saw a New Heaven and a New Earth, for the Old Heaven and the Old Earth had passed away." Everybody was getting the same information, that the Earth shifted her frequency on December 21, 2012, and people shifted from Doing into Being.

People in the sessions reported that when the energetic transformation happens you just have a knowing, it's not a belief, it's a knowing, that this is true, has always been true; it is a universal truth, it's not my mind making it up. That is what I experienced in Mount Shasta; it was a knowing. I had reunited with my I Am Presence and nothing would separate us again. Since that time we have been through a great cleansing. Karma from the third dimension ended on December 21 as the Earth moved out of dimension that has chronological time. So now we need to release all of the old emotions and negativity that would hold us in separation. A good way to do this is to use the ancient Hawaiian ceremony of Ho'opono pono where we ask forgiveness of everybody for anything we may have done which has harmed them in thought, word, or deed. We then forgive them and ourselves for anything they have done to us. It is essential that we release all old feelings of anger, resentment, self-denial, lack of self-love, grief both toward ourselves and others. All of our structures both internally and in the systems that support our outer lives such as politics and banking are now being realigned with principles of Oneness. Old systems must dissolve so that they can be recreated in harmony with the Law of One. So we will see much change over the next few years, but we must remember that everything is working for the return of the planet to Love.

Things are very difficult on Earth right now. A lot of people are in despair, and they can't see a future. They can't see where we are going. This is part of the plan, that we would fall farther and farther away from the light and the energy of God or Spirit or whatever you want to call it. This lifetime is the culmination of all of our lifetimes. This is the last lifetime for most of the people on this planet. The reason there are so many people on the planet is that nearly everyone who has incarnated here has come back for the grand finale. This is the last opportunity to repay karmic debts that we have accumulated in our whole cycle of incarnation. That is why many of us have had multiple relationships in this lifetime, to give partners of old the opportunity to pay their debts to us and vice versa. We don't want to have to come back on another planet in the third dimension to repay old debts from this one. As souls we choose to clear any energy of harm we have done by choosing to come back to make amends for the harm we did previously. Nobody judges us or makes us pay as a punishment. We choose for ourselves how to put right something we believed to be wrong.

We came to lift Mother Earth into her radiant light body, and when this is accomplished we will move on to other star systems and create new worlds.

Chapter 2

Coming Home to Yourself

*H*eather is a very gifted artist and musician. She channels beautiful mandalas from the higher dimensions, which hold keys and codes for wholeness and the evolution of humanity in a similar way to the crop circles. Heather channeled the mandalas for the front covers of both this book and my previous one, *Coming Home to Lemuria*. She remembers and speaks many of the ancient languages, including Lemurian. I guided her in hypnosis to an island where she could connect with her Higher Self, who is called Huulanaya. This is the name she uses for her website www.huulanaya.com. Heather has produced thousands of circles over the years and will be putting them into a book. Just seeing the circles helps one to connect with the other planes and the creation codes.

Heather—The first thing I see is a white star above the island. It is pure white with green, and it is incredibly balanced. I am walking to the center of this white star—it is a six-pointed star. It looks as if it is over snow-capped mountains.

Charmian—How is the energy of the star? What does it feel like?

H: It is from the land before and after time.

CH: Absorb the essence of yourself from this time in this dimension in this place. Feel yourself as if you are made from the light of this star. What does that feel like?

H: Pure freedom.

CH: Feel that Limitless Being that is you; this is who you are. What are the special qualities of this Being's energy?

H: She is a wand of pure light; she breathes magic. Pure light is coming from her hands and her feet. She bathes people with it. She is a Being of purity, grace, and elegance.

CH: Does she have a name?

H: Huulanaya.

CH: Bring her energy into Heather's body now. Why has Huulanaya chosen to embody on the Earth at this time?

H: She is a gift of Light, unheard of on the Earth, to penetrate the core of the heart. She understands the language of humanity, and she understands the distortions and frequencies that are causing disharmony in heartbeats. She is relating very directly to cancer and to the blood cells. She knows how to change this frequency.

CH: How does Huulanaya do that? How can she translate that knowledge onto the physical plane?

H: She must breathe through her book that she is creating directly with Mother Earth. She records the new eon in her breath even when she is sleeping and sings it into being.

CH: Is this a book that Heather needs to publish with her drawings and symbols?

H: Yes, it brings order from chaos.

CH: Let us have a look at that order emerging out of chaos. Let's go to December 21, 2012, and look at the New Earth that is being born.

H: It is done. All the forests are illumined in 2012. The forests are just naked and honest. Their sap, their Beings are bursting with intelligence for us. The trees are all in harmony with us.

CH: When the new frequency comes to Earth, how does it affect people?

H: They grow intuitively with the beat of nature and less from the heart of the mechanical computer world. The rhythm of nature is felt. It's like sign language; the energy of the invisible realms becomes so strong and supported. It's like somebody who has just learned sign language and is able to communicate. We communicate with our aura, our force fields, really efficiently. People begin to feel it without understanding it, but it is automatic. I am feeling the star in my throat because people are losing the power of vocal communication; it is being distorted.

CH: Imagine you are in December 2012 feeling the energy coming down. What does that feel like?

H: It is blissful, complete relief. The first thing I see is a fork-like shape connected with India.

CH: Do you mean like Shiva's trident?

H: Yes, that's what I am seeing.

CH: That represents the forces of destruction and rebirth.

H: There is a massive celebration of the vitality of life. My body feels entirely blissful; it's as if the prayers of the people of the Earth are filling my body. Huulanaya has written of this; our Prayer Bodies are lightening up. In India art is life. Like in the story of your near-death experience that you just shared on You Tube, every breath is precious and by 2012 it is gathering momentum. The Earth is becoming the incredible athlete who will win the race. She has been training, and there is nothing that can stop her. The White Star is connected to the absolute center of the Earth where prayers are gathering from the invisible realms as well as the human realms. We are breath; it is everything. One of Huulanaya's teachings is Miha, sharing breath. People are losing the understanding of the importance of breath. They don't know how to pray, and they are frightened. They need to know that just being here and breathing is all that is asked for. It's like leaves—they give oxygen, they don't need to sit praying out loud or to know how to pray. People need to know that their breath is really sacred.

CH: They need to know that every breath is a prayer?

H: Every breath is a prayer in silence. Breath is understood by all Beings. I am seeing it is like the dolphins when they squeal; they don't need to be understood. Each one who receives it decodes their message. Some people understand it, and some people know how to pray correctly, but breath is enough. The more that people breathe without expectation, the more they will receive the wisdom of the trees. People who need help can receive it from the trees—particularly the children and the teenagers in places like London and the inner cities. They are so busy with their minds, so many questions. "What are we going to do next? What job shall I do?" People are confused, they don't know the answers, and it is causing suicide and all sorts of things. People don't know what is around the corner. If they can just focus on the breath, every single breath can have an answer. By December 2012 we are within the harmonies of the new paradigm.

I am being shown the Golden Light that you traveled to in your near-death experience with a cross in front of it. It takes me back to a memory of a time when I sat in a cathedral in Glasgow. I sat under a picture of a pure white dove, and I wept and wept and felt so heavy. I realize that it was a message of what was to come. It's like we had to go through this time of immense density. The symbol of the dove is so important; it represents our purity. This is why visions, color, and sound are so important because people are losing faith in words. The stronger the image, the more it speaks to the soul of each person so they don't feel dictated to. This has caused much disruption and rebellion, because if you are told what to do you battle against yourself. Whereas color, sound, and vision, dolphins, whales, and nature are there, balancing it all for us. All the metal is oppressive, from cars to tanks and factories. There is a natural harmony being held by nature and trees.

CH: After December 2012 what happened to the world of metal and factories? Do you see the new technology that replace gasoline and other such fuels?

H: I see crystal chambers, all of our auric spheres and galactic entrepreneurs at work.

CH: Do you see the Star People?

H: Yes, extraterrestrial Lights helped me in my pregnancy. I am being shown some of the circles I have drawn. People are being guided. People are meeting and, as the breath gets stronger, there is an image of harmonizing. So when you meet people, don't rush; share breath together like the Hawaiians and Maoris do. It is something to do with the breath and the third eye connecting. (In

Hawaii when people meet they put their foreheads together and breathe together, sharing "Ha" or breath of life.) When you meet like this, you bond energetically, sharing breath and Mana (soul food). In the future we are actually in one frequency, all connected, and what we breathe is the Absolute. The space for argument will begin to dissipate. Peace is held sovereign, and the Truth will become so regal that it is all that is recognized.

CH: What happened to the chaotic energy of earlier in 2012? How did that shift?

H: I am seeing crystals planted in the Earth and skulls.

CH: Do you mean Crystal Skulls?

H: Yes. Source can infuse immense power into the human skull; it is linked with the crop circles, it is galactic. Bicarbonate of soda on the tongue propels the heritage of negativity very quickly.

CH: You just put a little bicarbonate of soda on the tongue?

H: Yes, or in bathwater. All of the old "stuff" is gone, no matter what it was. People are carrying guilt for stuff they have done, whatever it is. Bicarbonate of soda will release it very quickly. People are speaking from their own history, and others are judging them from their own history. Bicarbonate of soda will shift it all.

CH: Can you tell me about the connection between the children and the New Earth?

H: Breathe with them; they are breathing from the Oneness, and you just need to look at the palms of their hands. They carry the new information just by being. If people could be more patient to breathe around them then they would receive their energy. They are of the purest frequency we can behold. We have an immaculate human form already in the future, and through the breath all obstructions can be dissolved. The children carry the new energy already in their breath and in their heartbeat. Their words are forming with ours, as is their language. If we could see them in their energetic fields, they are like unicorns coming back. They are so beautiful and so pure. For people who don't believe in magic anymore these children are bringing it back. They are the link. There are babies being born now from eternity.

I am seeing the softest blues and violets. Many, because of the catastrophies, the wars, and the news, have distorted the word "angel"; a lot of people just don't believe in anything anymore. These children are carrying pure unconditional love. With advanced Kirlian photography, which records the aura, you can see the effect people's colors have on one another. These children touch us with their pure energy. They carry time in a different way.

I am being shown that many people would give anything to be able to swim with or to be near dolphins. If they understood that these new children have the same quality, you just need to be close to them. You don't have to touch them to receive their energy.

CH: We receive their energy just by being in their aura?

H: Yes, and if you receive eye contact you are lucky. Silence is so important. In their silence they tell you everything, energetically. One day there will be recordings of what happens in the energy fields when an adult meets one of these new children.

CH: The word I hear is communion, an energetic merging in Oneness.

CH: Can you tell me about our relationship with animals after the shift?

H: They own us already, we are already One. By thinking of an animal you are one with it and it with you.

CH: What about our pets? How are they helping us?

H: We are lost in time; we are mixed up in the wrong dimensions. The animals help us to stay grounded and present. They are helping us to remove animosity; they are teaching us how sacred our bodies are. The six-pointed star is to be placed in the minds of all humans and all animals. It is in the Earth. It is uniting us all.

CH: What does the six-pointed star represent?

H: Peace, love, and truth.

CH: Where does the star come from?

H: Sirius.

CH: Is it connected with the Christ energy—the Christ flame?

H: Yes, the Christos. Beyond all Light is an energy that has not been voiced yet.

CH: Let us invite the Sirian Council to be present with us and ask if they have a message for us at this powerful time on the Earth.

H: Much pain has been dissolved. The salts of the Earth have been awakened to infuse harmony through the blood. It is bringing the Christ element and is birthing a new form on the Earth. Understanding of the children as the people of now and forever—the silence—the natural sensitivities of our Being are to be realized in 2012. The Grace into which we were birthed will be reawakened.

Princess Diana held many teachings for us all on how we can dissolve the pretense that we can live in harmony with our wholeness while we abuse the symphony of the Earth. People are lost in the war zones, but they will return to the Christos. We have not infused poverty for any other reason than to teach luminosity. We have a very high frequency in Sirius that will engulf the heartstrings of every soul on the Earth. Everybody will feel connected to the magic that is infused in every breath and with many synchronicities. There is a field of health on the Earth, and when this field is felt it will nurture the silence and it will awaken the symphony of sound that we know we were born to hear and to receive.

When we are babies we Are. We are not here to destroy or to speak ill will in any way. We are here to birth innocence and peace. This is what we emanate in our breath. When you are near a child this is the emanation you receive. You receive the gift of their Being in their silence and their purity. Never hold a child in anything other than Love. People are empty and must be filled with Love. When you touch, you love. May every flower be remembered as a smile.

CH: How can people connect with Sirius now?

H: Be thankful every night for the day just gone. Remember the gift of the day and take it into the dreamtime with you, then awaken the next day in gratitude—not in fear. We don't have to try to be anything; we already are. Someone may stand before us appearing to be angelic, but the person looking at the angel is equally angelic without needing to understand or comprehend. Sirians are very real, and in order to be real we breathe. Cancer will be diffused through Light, a pale green ray. Many hearts are conditioned with the old way.

CH: What is the old way?

H: Old teachings. Much has to do with food—McDonald's and supermarkets. The new frequencies will guide the Hands of Light toward health. People will grow toward health. People will see colors and know how to transcend limitations. We are moving through it with every breath.

CH: People are already feeling drawn to eat more live and healthier food.

H: Remember, with every mouthful you take, to breathe harmonious balance to the world, to those who know nothing. There are many who may only be offered a piece of spam on a plate because that is all there is but may that spam on the plate be as sacred as a walnut. Make all eating a sacred experience. Bless everything.

New magnetic forces are being born so we will be attracted to that which will serve the highest. The unicorn's horn will never leave us. The unicorn is a symbol of the Divine Light in which we were born. People lose faith because they think they can never be what they are, and yet they already are this Light. They have chosen to walk the path of difficulty, and this path will take them home. Through color, sound, and glyphs the remembering will happen. People's eyes will always share the Truth. One day nobody will be able to lie. The Truth will be anointed for us all.

CH: How can we help those people who are in fear of the changes that may come, as we go through the shift such as tsunamis and earthquakes?

H: I am seeing the Earth coated in color. Remember the Buddhist chants, remember that we awaken with Love; remember that our bodies are growing with the Earth. Celebrate the body no matter what conditions we have. By loving our own bodies, and being in harmony with them, we will understand how the Native Americans never separated themselves from the Earth. There is an old saying, that even the lightness of one feather can help a bird to fly. Whatever we do to help the Earth—it may just be picking up a can by the side of the road—as we become more and more harmonized with the Earth we bring the challenges to a halt. We all become co-creative in the balancing, and as we breathe tornadoes into ourselves they become part of us. We allow them to settle in any part of our body where we feel disease or we do not feel in harmony with. We must recognize even the disease as part of the Earth. The more we take care of our bodies and ourselves, the more we are connecting to the Earth.

Plurals are so important—we—the communities that are evolving, being together. The children need to know they are not alone; friends surround even a blade of grass. Do not be frightened of the Earth. Do not be afraid of the universe, because she is teaching us. I remember in Hawaii at Ho'o kenna beach, there was a little boy who had fallen and had blood pouring down his face. I went over to help clean it up and put a band-aid on it, but his father came over and said, "No, he's fine." He said to the boy, "Go get in the water, boy, the rocks are teaching you." This is the Oneness with the Earth. Know that balance is coming, the indigenous voice is being heard in the ethers and the codes are being implanted and we are receiving them. As we clear our frequencies (soda bicarbonate helps with this), we will unite much quicker with the astral plane. Many people are oblivious to all of this, but it is innate within us all and the Earth will settle as we settle more into the Truth of Who We Are. It will take time, and if we could remember that our blood is like lava, the speed of the blood will slow down and we will all become more rooted again as family, one big family, a Rainbow Family.

CH: Is this after the shift?

H: I am seeing 2018—an incredible whole feeling. Our families must not forget to be positive. There is too much chat around tables now that is of fear.

CH: People have lost the connection to the Greater Plan. What can we tell them to help them to move through the fear?

H: That everything has consciousness, and as we learn to communicate with the elementals they will reflect back something more nurturing.

CH: Do you mean that as people begin to find peace within, it will be reflected in peace without?

H: People are just speaking from ignorance, but there are devas, plant spirits, angels, and the nature of our spiritual beings that are evolving. All these will help to counteract what is coming out so these negative conversations will be halted.

CH: This will all just happen as the energy changes on the Earth. People will become more positive?

H: Yes. I am experiencing a pulse, waves of color, silence. We were born of these frequencies. Do not want for things. Do not be wanting just Being. Being allows What Is to emerge. Trusting—the elementals have invited us here and we have forgotten that we were invited. A mushroom was created for us and yet we pick it without consciousness. The native people always thank the spirit and give a gift before they pick something or kill an animal for food. They make an offering to the Spirit of the animal, plant, or tree. They are trying to awaken the union with nature and with Divine miracles. It is like your teaching when you trust that everything is already there and has already happened. As you say, the more we can realize that everything is here for us, then we can be at peace and peace will reign on the Earth. It is a gift of unity between two worlds, as the world beyond influences the world within. There is a harmonic tone that can help us to evolve into Divine consciousness. There is too much striving at the moment.

CH: That's right, we don't need to strive at all.

H: That is why people must remember to breathe.

CH: From striving comes strife and from breathing comes life.

H: The whales are helping to engulf the planet with the new drug. It is just Love; they are breathing it and singing it into the water. When people can start to slow down, when we get it, when we shift, then we realize that the breath is like a Love drug. We are 70 percent water and the Love is already within us. We don't need the drugs to connect to the Love.

CH: No, we need connection. That is all. That's why people take drugs; they really want connection. They need to be shown a different way to find it.

H: Mother Mary is in a different language, a flower is of that language, so is a tribal elder and so is a chipmunk.

CH: Is there any last message from Huulanaya in completion?

H: Mothers must be held above all. Mothers must be engulfed with supremacy.

Chapter 3

The Galactic Alignment of December 21, 2012

D was visiting Hawaii when we did the session. D is an ancient Lemurian soul who was called to return to the Hawaiian Islands to have his Lemurian light codes and memories reactivated. He first went to Kauai where the energy of Lemuria is very strong, very present, and clear. He then traveled to Maui, which is also a very powerful vortex of Lemurian energy. The Hawaiian Islands are remnants of the vast continent of Lemuria. Since Harmonic Convergence in August 1987, the Hawaiian Islands have played a critical role in reversing the negative effects of Humanity's Fall from Oneness and in the healing of Mother Earth. Many of the people embodied on Earth at this time are Lemurians returned. They realize that they are here now to heal the atrocities that resulted on that continent when Humanity fell into the abyss of separation and duality.

D is a somewhat isolated individual who is not comfortable with social contact but who has a very good connection with the other dimensional realms. He was able to access information about the alignment with the galactic center and the effect that it will have upon the transformation of the Earth. He saw a black hole in the center of our galaxy that shifted the whole frequency of the Earth when it came into full alignment with it on December 21, 2012.

Charmian—The great cycle that began in Lemuria is now coming to completion. We journeyed from the light through the darkness, and now we are returning to the light. We who are the family of light, who have been walking upon the body of Mother Gaia for eons of time, have come now to the final days when the Earth shall be lifted up into her eternal body of light. The children of the light who have been carrying the frequency of Oneness, which they brought to the Earth in Lemuria, are now being positioned so that they too may

16

make the ascension into their radiant light body. This is why many are called to the ancient islands of Hawaii, which were part of the body of Lemuria. The land remembers, the land holds the frequencies of the Oneness.

It is very important that you (D) have returned to the place of your origin to receive the codes for the final and complete activation of your Ascension body. This is why it was necessary for you to be on Kauai, for this island most closely resonates with the vibration of your soul. Your body in Lemuria was much less dense and physical than the body you are in now at this time on the Earth.

Let us go now to the Lemurian temple on Kauai. What do you see?

D: I see a light. It's almost like this radiant light coming up from the sea.

CH: The source of the radiant light is the Lemurian and crystalline city where you lived and worked. In front of you is a crystal city that emanates a beautiful radiant light. Tell me what you see.

D: I see a golden arch. I see two pillars on either side of that arch that seem to reach up forever. I see a pool of liquid light or mercury. I see Beings all around the pool and streams of light. Gazing into the pool I see infinity. It is deep and there is a palace. The palace is translucent; there is a huge jewel like an emerald hanging over the gate into the city and over the temple, which is made of light.

CH: Connect with the Beings who have gathered around the pool. What is the purpose of this gathering?

D: They are performing healing; it is for healing the Earth, Gaia. It is opening up a portal to God, to the Divine light, so she can be healed.

CH: Who are these beings who have the ability to do this healing?

D: They are priests, Beings of light, Beings of pure love.

CH: Where did they come from?

17

D: They come from the light; they have always been here; they are everywhere and from everything. There is no time or place in particular. They are here to serve.

CH: Are you one of these beings?

D: I am being taught and ordained.

CH: What is the mission you are being ordained for?

D: To teach others, teach them to serve and to interface with the Divine light.

CH: Why has your soul brought this memory to you now? Why did you need to revisit Kauai and to remember Lemuria now?

D: It is important for me to remember because I am asking to remember, to feel, and to know what it feels like to be supported and loved, to be connected and nurtured, to be inspired and awakened. I have been asking to be shown my purpose and to be given guidance for this journey.

CH: Are any of these Lemurian Light Beings here now in physical form on the Earth, or are they always in light bodies?

D: They can come and go at will, they don't stay, they are mostly in their light bodies, and they can take any form at any time as needed.

CH: Let us bring that aspect of you as a Lemurian priest forward now at this physical time and see his energy merging into yours. (Pause)

CH: Let's look now at that time when the Earth and the human race are ascending into their light bodies. Let's move to that specific time where the Earth is going through her energy shift and the children of light are going through this shift with her. She is moving from the third dimension into the higher frequencies of Oneness. It's like a golden doorway, and when you step through it the energy has shifted completely. Tell me what that looks like to you.

D: I see that eventually it looks like the Earth being sucked through a black hole, a vortex, and it's almost like the Earth turning inside out. I see a lot of light. I see a lot of blues and greens, a bluish green. It's like the Earth will be surrounded by this grid of light, and it's almost like the magnetic field of the sun surrounding the Earth and elevating its frequency and vibration.

CH: On December 21, 2012, the Earth aligned with the black hole in the galactic center. Was this the event that will open this vortex?

D: I see the vortex is opening, it's shifting, it's already shifting. The vortex has been opening and then at the end of 2012 it's like this big ring that the Earth went through. The black hole has an accretion disk, and once you go through the disk there is no turning back, you get sucked through. The Earth reached that place in December 2012. It reached the black hole, then it will reach the point of no return and it can't go back, that was the point of transformation when we went through the black hole.

CH: What effect did this event have on the people of Earth?

D: Individually, collectively, people are being called to remember where they came from. I get a sense that this is a time for everyone to choose their path, to choose the direction they are going, to choose to follow their destiny, their soul, their spirit, their calling. They need to remember, to awaken, or to be lost. After the end of that year as we went through the vortex there are many who were lost.

CH: What happened to them?

D: They will have to keep coming back to Earth again and again. Those that are awakening can ascend and transform and don't need to return. They pass through the portal, the opening into the Light.

CH: The ones who return to the Earth, do they leave their bodies at that time?

D: No, I don't see that.

CH: Do they carry on being on the Earth the way it is now?

D: That's what I see; they will create their own suffering and their own hell on Earth through their own ignorance until the next cycle, the next opening. I see a time when the Earth will shift all at once. I see that the whole Earth will shift, and everything and everyone on it will shift. I don't see that it's in this cycle; I see a lot of people returning.

CH: Let's look at the people who do go through the black hole, the vortex. How does life look for them?

D: I am seeing the Earth as having free will, freedom, free expansion to be able to move between dimensions, realities, universes, and parallel universes at will. All Beings here will know the Truth, know God as One, know Oneness and be able to travel interdimensionally, intergalactically with the sense of the connectedness of All Things. We will have had an inner transformation with no need to return to embodiment anymore.

CH: So are we all in our light bodies?

D: Yes.

CH: Are we still living in communities like we are in now on the Earth?

D: I see groups of healers and schools, Ascended Beings, and Masters. There is always work to be done because it is not just on Earth but also in other dimensions and other galaxies that we help other Beings.

CH: What does D need to do to prepare for the ascension?

D: Spend a lot of time in nature, the elements, connecting with nature outdoors. I need to be in community, studying with like-minded people. I need to study with those who are working with the vibration of the Ascension, those who are moving in the same direction. I need to be with others, not to be alone.

CH: Is there anything that people in general need to be doing to prepare for the shift?

D: My sense is that people want communities and connections based on togetherness. We come together as one. None of us can be an island; we need to support each other. It's either all or none. We need to try to bring others along, learn to love all aspects of our divinity and to embrace everyone. Embrace the dark and the light, all Beings. Give space to everyone to unfold as they choose. Let us not see anyone as separate from the light, not separate from that which is the same, seeing everything and everyone as what can only be love and light.

CH: Can you give information about Charmian's role?

D: I see you wrapping yourself in your arms, wrapping yourself around like a shawl, wrapping yourself in the shawl of Divine Mother love. I see this shawl is shaped like a heart, and I see the heart transform into wings. I see the wings of an eagle, purity of light, and not to be afraid to fly. You can soar and allow yourself to flap those wings. Don't be afraid to take off and fly from shore to shore. Wherever you land, you will be supported. Wherever you go, you will land on your feet not on your head, so don't worry.

CH: What message do you have for D at this time in his journey?

D: Trust, just simply trust, let go and allow more just to be. Allow things just to flow when you encounter another person. Remember everything is in the Divine flow or vibration and it's all OK. Everything will unfold. Trust that it is all for the highest good.

CH: I feel there is something from the temple in Kauai that you need to carry back with you when you leave Hawaii. What is that?

D: Compassion for myself, something higher, a higher calling, being empowered. I need to be able to be empowered, to hear, to truly understand, and to know that I can hear clearly, that I can understand and know the truth within me. To know that I can see (psychically), can feel connected, and to know that I have a loving heart and a truly important mission.

CH: What wants to come in is an emerald tablet. It needs to come into your heart, and it will attract others to your vibration. So let yourself feel that green emerald coming into your heart and into your whole body and know that you are love and you are loved.

Chapter 4

Lifetimes of Preparation for Our Return to Oneness

I had a hypnosis session myself to look at the New Earth and how I had prepared for it in many of my lifetimes. I have always been a healer and a priestess working in the great temples of Light from the time of Lemuria, when we lived in Oneness, through The Fall, when we chose to experience separation from the Source, as well as all of the civilizations since then—Atlantis, Greece, Egypt, Mayan, Incan, Druid, India, Africa, and Australia. I have returned time after time to train men and women to reconnect to the Source, to learn how to use that connection to heal the Earth and her people, and to serve The One in ceremony and devotion. Many of my students in this lifetime are those whom I trained in the ancient Mystery Schools. We have had all the training we need from many lifetimes of service in the temples, and we are well prepared to make this great transformation that is happening on the Earth just now. The therapist I was working with (I have called her E) guided me to a lifetime that was important as a preparation for the time we are in now on the Earth. I was taken to Ancient Greece and the temple of Aphrodite. From there we traveled to an existence on another planet in another universe that is directly related to what we are experiencing now on the Earth. Many people who are here as Lightworkers on this planet are actually Beings from many star systems who have come to help the Earth to evolve to her next level as a fifth-dimensional planet.

E—Where are you?

Charmian—I am in the temple of Aphrodite. Priests of Apollo are working with the fire and the sun. Priestesses of Aphrodite are working with the water. We have a pool and a spring. We sing, we bless the water, we call down Aphrodite, we bathe, and we take people into the pool. We give the blessed water to people to drink and to bless the babies. People bring their babies, and we take them into the pool. We ask the Goddess Aphrodite to watch over them.

E: Go to an important day in this life.

CH: We are leaving. I am with my twin flame at the beginning of the Greek era when the energies were very pure. My twin flame and I decide that we have finished our work here and it is time to go Home.

E: So you are leaving the physical body?

CH: Yes, everybody is there; everybody has come to say goodbye and to help. They have drums and trumpets, crystals and crystal bowls. We lie down and hold hands. We have our golden sun disks over our hearts. People are playing music and chanting. A light comes down, and it focuses through the disk. Our bodies are getting very hot, but it is very joyful.

E: So is it a conscious choice to leave?

CH: Yes, we are glad to be going because we have done good work, we have helped many people, we have served the Earth, and we want to go Home now.

E: Tell me about the process of going.

CH: One minute I am in my body and I am hearing the music. This light comes down and it has come from Home (sobbing). It has come to bring us Home. It is the golden light that we live from and it is everything that we are, it is Home. It is everything, it is ecstasy, it is so hard to be away from it; it is just ecstasy.

E: Is your partner with you still?

CH: Yes, we are drifting up, we have left our bodies, and they incinerate themselves like Jesus did in the tomb; they just combust and disintegrate. We are just floating up this golden pathway, and we are still holding hands. We just want to go together. We want to be together because at other times we have not been, because we were separated—not by choice. When Atlantis went down, we were separated. So this time we said we are going together and we are going to choose the time, so that is what we did.

E: Can you describe what Home feels like?

CH: It's just Light, just golden Light, and we merge into the Light so we don't need bodies anymore; we are just consciousness. We can touch each other's consciousness all the time. We are never alone, never lonely, just a joyful weaving in and out of this consciousness and that consciousness. I feel my best friend from England who died recently is here; she is very joyful.

E: She is Home with you?

CH: Yes, she is waiting for me. It's like a joyful reunion. My two children are here also.

E: As you look back at that life as a priestess of Aphrodite, what did you learn from it?

CH: The joy of serving; even being in the body you can have joy. You don't have to wait until you go Home. If you are in circumstances that are harmonious and where you can fulfill your soul's purpose, it doesn't have to be painful, it can be joyful. Many of my soul family were with me in that lifetime; almost the whole group came down. You don't come alone; you bring your family with you. It's just a matter of how many choose to come for each life. In that lifetime most of them chose to come, and in this lifetime they are all here, the same ones.

E: Once you go Home, how do you go about deciding where to go next? Do you have help?

CH: We all get together, nobody decides alone, and there are other consciousnesses there that know what is needed and we know what is needed. For instance, certain parts of the universes, the created worlds, may need help so they put in a call and we decide how and who is going to answer that call. We decide what form we want to go in. It might be a tree or a human, a crystal or a dolphin—it's all One.

E: So it depends on where help is needed, where you can serve your purpose. Take me to your next journey. You have decided along with your soul family and the other consciousnesses where you are going next. Can you tell me what you see?

CH: We are coming to this planet that seems to be all water. It's all blue; it's surrounded by a beautiful blue aura. It's

24

not so much physical water, more liquid light. It is a very high frequency. This is the Blue Star I have sometimes journeyed to. It is very lovely, very peaceful and harmonious.

E: Do you have a physical form?

CH: No.

E: Do you have an energy body?

CH: It's more like an angel or a bird. It has the wings, but not like a physical bird.

E: Can you observe the planet?

CH: It's just all blue light and there are Beings in the light.

E: Do you recognize these people?

CH: Yes.

E: Are they part of your soul family?

CH: Yes, but they seem to stay here, and we come and go a lot. They look different. They are very tall and thin with no hair, big eyes, and long fingers. They are very ethereal rather than physical, but you can just see a shape. They are not very dense. They have incredible blue eyes, faceted kind of eyes that shine like diamonds. They are so kind and loving. They have evolved so far that they are just loving and peaceful and nothing ruffles or ripples them. They are calm and flowing.

E: How do you communicate with them?

CH: By telepathy.

E: How do they sustain themselves?

CH: Just from Light.

E: Is there a name for this place?

CH: It's just called the Blue Planet.

E: Is it in this galaxy?

CH: No, it's not in this universe. To get there you go through the middle star in the belt of Orion. It's like another universe, another dimension; it's a gateway. I have been there before, but they are telling me that our soul group went to the Blue Star as a preparation for coming to Earth because the planet we are coming to, the Earth, is very chaotic. It was already known that all of this was going to happen, The Fall, all the density and the darkness, everything. So we decided to come here first because these Beings are so calm and timeless. When we came to the Blue Star, they gave us a seeding, like a DNA coding so that when we came to Earth we would all have that. It is coming out more now because our bodies are becoming more crystalline. It seems like the DNA is also becoming more crystalline; it's almost like they gave us a microchip so that we could tune into them and they could travel here. It's like a homing device. So these are ancient wise beings; they have so much wisdom and so much love and they want to help. They are here to help, and because they have this connection to us through the microchip in the DNA that gives them a way to communicate with us.

E: Are there many beings on Earth that have connection with these beings from the Blue Star?

CH: No, just a few.

E: When these beings like yourself come to Earth, how are you able to help?

CH: We help everything because of the consciousness that We Are. We are just this Light and this Love. We help the Earth and the water.

E: Just with your energy?

CH: Yes, just by being this frequency. It's like the energy of pure Light and pure Love. We don't have to do anything but just glow. When we go into the water, it goes out into the oceans, and when we walk in the fields, it goes out to the plants. We feed everything with our Light and our Love. It's very joyful and wonderful to be able to serve the planet in this way. We are very grateful.

E: We seem to be coming to a crossroad or a very important time of change and transition on the Earth, not just for humanity but for the Earth as well. Can you tell us about that?

CH: This is the time that we came for because we have this wonderful blue energy; it's like the Christ Light, and it's quite a shock coming to Earth. It feels really strange, it's hard, it's heavy, it's like, "Why am I doing this? How did I get in here? Let me out." So this is the test really, it is to become the Light in a physical body. This is the test.

E: Have you had many Earth experiences?

CH: Yes, lots of times—always coming back, many seedings. We go and we come, we go and we come. It has quite dramatically failed lots of times, then we have to go to regroup, to try a different way, and to refine. We are always refining, trying to get a better fit. Bringing Light into the physical body hasn't worked very well. We didn't get it right.

E: Are you speaking specifically about your Light?

CH: Everybody's. We didn't realize how density would affect consciousness. We thought we could maintain the purity and the integrity of our Beingness here, but we couldn't. Being in a dense body seems to be such a pull and a distortion that people get sucked away from the Source, the Essence. All the power and the things glamorize them. They are pulled by the delights of the senses, and they forget about God.

E: They forget about anything beyond the 3-D experience?

CH: The more they go into the 3-D, the farther away they get from the Source.

E: Is there anything they can do to help them to return to their Essence?

CH: Remember Love, remember the Love. It is happening and that is why so much is disintegrating now. We have pushed it so far away that it is crumbling at the edges; the foundation that we have built this artificial world upon—the very foundation—is crumbling. People are trying to hold on to the pieces, but it needs to fall so they can remember that they can lose the outer and it just makes the inner more visible. You can lose your "stuff," but you can never lose the connection to the Light. It's just that some people are farther away than others. The Light is in everybody, but some people have got farther away from the Source.

E: Is this your purpose when you come to Earth, to help people to remember who they are?

CH: Yes, because I have always remembered, I never forgot who I was.

E: Does that make it hard for you because you do remember; you know how wonderful it is and how different it is?

CH: Yes, I feel I have been waiting for a very long time. I feel tired. I've been here a long time waiting and having a lot of very painful experiences, really taking on human emotions, understanding human pain and suffering. I know all about that; I have a PhD in human suffering. I came to show them the way back—that you can go through the pain and it doesn't become who you are. It's just something you pass through.

E: I would like to invite Charmian's Higher Self to come in now. Why did you show Charmian the memory of the priestess life?

CH: She has had so many lifetimes where it hasn't been good, where she has been a priestess or a seer or a psychic and she has been burned or tortured and put to death. She has been afraid to step out and be who she is because of the many times she has been killed for being herself. She needed to see that what she knows within herself (crying)—the way to be, to serve—what she knows to be true that she has actually lived, that it is true and it will be true again. In this lifetime it will be true again.

E: What is the Divine purpose of her relationship with her twin flame?

CH: They serve the Divine plan by embodying the masculine and feminine aspects of God and by putting that energy into service through sacred union—the sacred marriage on all levels of Being.

E: Charmian seems to have some difficulty about being separated from Home.

CH: Home is coming here. The New Earth will be like this one but much more joyful, and she won't feel the need to be always wanting to go Home. She will be Home already and so will everybody else. All of the people that she loves will be Home, they are all there. It is paradise, a wonderful world. The cats are there.

E: Has she crossed the marker for Ascension?

CH: A long time ago. When she went Home in the near-death experience and was given all the codes. They said that this is why she came back. They asked her to bring the codes back, for the Earth, for the water, for the trees, for everything. She has been carrying them all this time, wherever she goes. She has traveled a lot, and in every place she has been seeding the codes.

E: Has this been important?

CH: Oh, yes, it has transformed her life.

E: Charmian has been using crystal singing bowls in her healing work. Is this part of the Ascension preparation?

CH: That is what they used in Atlantis and Lemuria. The crystal sound breaks up old patterns; it harmonizes systems and energies that are out of balance. The sound brings them all into harmony. The sound tones work on physical, emotional, mental, and spiritual bodies both in people and in the Earth. They can be used for combining sound and intention; the bowls could be played on the beach with the intention to heal the water. You could play them with a group of people all holding the intention for the water on the Earth or the air to be healed. This is very important, as pure sound is what is creating the new world. The sound is the root element that life is created from. In the beginning was the word and the word was sound. So for anything that you want to bring into manifestation use sound and intention.

E: Charmian feels a strong connection with Sirius. What is the connection between Sirius and the New Earth?

CH: It is her second home. She feels so much love coming from Sirius. She feels completely loved and supported there that she does not always feel on the Earth. That is part of her problem; she gets so much deep unconditional love, acknowledgment, and appreciation from all the other planes because they see her Light and how hard it is for her to be in a physical body. They really encourage and support her. They see the dedication that she has, and they know her purpose here so they want to help her by doing healing on her body to make it easier. They are doing it now.

The trick is not to leave the body and go Home, it is to make the body lighter then there will be no pain. The way to do that is by focusing each day on doing what brings you joy and intending; "I'm only doings things now that bring me joy." It doesn't matter what it is. If there is a situation that is not joyful, just leave and walk away. Say to yourself, "I don't need that right now, thank you very much, goodbye."

When you get triggered, know that it is something in you wanting to be healed; don't run away from the situation, look at what is asking to be healed. You may need to ask somebody to help you to access the wounded part of you that is asking for attention. Instead of running a conflict to the death, just say, "I'm triggered, thank you. I'll go and get help now." Then find somebody to help you— the therapist, a friend, or a counselor.

E: Do you have one last message for Charmian?

CH: Stop worrying. Know that every day is a gift. It's meant to be joyful and fulfilling and fun.

Chapter 5

Earth's Final Clearing from Energies of Control & Resistance

In the previous chapter I looked at the lifetimes I had had on the Earth in preparation for the Dimensional Shift and the return to Oneness. I wanted to have personal experience of the shift itself so I had a phone session with my good friend Colin. We went through the dimensional doorway and looked at the New Earth on the other side. He and I have a very close soul kinship, and we seem to be able to activate each other's codes as every new portal opens up the next frequency on the Earth. We journeyed to the December 21, 2012, doorway by first going farther into the future and then traveling back to see how we got there. We created a time travel ship several years ago, and this is how we traveled forward in time to the New Earth after the shift. We constructed the ship with our minds in conscious co-creation with the intelligence and consciousness of the ship. This is something that will be very much part of our everyday experience after the shift. We set off in the ship and waited to see which time zone was calling us to enter. In my previous book, *Coming Home to Lemuria*, I described the crystal light cities that we created with our minds in co-creation with seed crystals. We then used the temples to bring our world into manifestation. These temples are now being reactivated on the planet as the frequencies of Oneness return to the Earth.

Charmian—We are up in the ship. We have gone out of time, we are traveling through space, but it feels timeless. We are coming to the Earth in the year 2035.

Colin—It feels like we are coming into one of the crystal temples that are now very well established on the Earth.

CH: Do you mean a Lemurian temple in one of the crystal light cities?

C: Yes, one of the very bright ones. We are already there when we arrive from the past; it's like we are meeting ourselves. We are both very light, we have made the transition from density, and now we are holding the Light.

CH: I can see a lot of control panels, and we are standing in front of them. It seems like there is still a lot of adjustment and fine tuning going on. Every year the light gets stronger and the density fades away, but there are still residual elements that need to be fine tuned. In fact, there is a feeling of heaviness. There is a resistance, and it is getting harder and harder to be around it or for it to be on the Earth.

C: It feels like the very darkest energy hasn't quite shifted yet. There is just a tiny residue here and there, but in those places the energy is very dark.

CH: It feels very dark and dense because the rest of the Earth is so light now. The contrast is much more marked, and there seem to be little pockets of people who are very resistant and are trying to hold onto the old ways. They are doing all this plotting and planning, having secret meetings.

C: The interesting thing is that we can see what they are doing quite clearly. They think they can't be seen, but we can see them clearly from where we are standing.

CH: In fact they are in a goldfish bowl. They think they are invisible, but actually a transparent bubble-like glass surrounds them.

C: It's almost like we are offering them the opportunity: "Do you want to play this out some more, or would you like to come into the Light now? Either way, you can't do this here anymore." It is still their choice, though, but we seem to be able to observe it now without getting triggered, we feel only love really. I feel more love for them than I have ever felt.

CH: We have moved so high up into ecstasy that is how we live now. We are living at this plane of ecstatic love and bliss, so it is easier to be in compassion for them because they are missing this experience, and they are not affecting anything else on the Earth now except themselves. It seems like they are in bubbles, they have been disconnected from Earth's energy systems, and they don't actually have any power, influence, or affect any more. They just think they do. It's as if they have been quarantined. I am seeing these little pockets, like little black bubbles all over the Earth. It's not that many of them actually. Some of them are underground like in bunkers. It seems to be that some of our work now is to lift these bubbles into the Light. It's why we are still here.

C: In fact that is why we dropped into the Earth at this time of 2035 to see how we are doing that. That is what we are going to do next, just ease the bubbles out and set them free.

C: It seems we either lift them into the Light or lift them off the Earth, somehow.

C: They get the choice.

CH: Yes, but they can't be here anymore. It feels like Mother Earth has decreed that they can't be here anymore. They have been given this grace period, but now it has run out. They must choose now, either to come and fly with us or leave the planet.

C: It feels like quite an emotional moment.

CH: Yes, I am very aware that they are us and we are them, it's not anything separate. It feels like a little brother has gone off the rails, and we want to lovingly guide them back. If they don't want to be guided, they need to move on. There are other Beings coming to collect them, that is the ones who are choosing not to come into the Light. Beings are lifting the bubbles and transporting them to completely different star systems where they can carry on that level of experience if they want to. It seems a shame because you don't know why anybody would want to do that, but it is entirely their choice.

C: Perhaps they still have some things to finish off or still to learn from third-dimensional existence.

CH: Anyway, that energy is clearing now from the planet.

C: It feels very clean now that energy has gone as if we have moved into another phase. We have stepped into a different energy.

CH: Let us look at what is happening on the rest of the Earth. It looks like fairyland, everything is shining, golden, and rainbow colors, it's as if everything is in Technicolor and 3D.

C: I can see people's hearts shining more than I can see their outline.

CH: Everybody is more ethereal and translucent. Their hearts are shining with Light, very brightly.

C: As you notice them and they notice you, there is an instant heart connection.

CH: When you see them, their eyes light up, anybody and everybody. They greet you as if they have been waiting for you all their lives. It is lovely.

C: It is so peaceful; you can feel their energy expanding your own heart.

CH: I am very aware of the children. There are still children, and they are very pure. That is one of the reasons why the black energy had to leave because the children need pure, clean energy. They are such clear, bright colors. The children are even more ethereal than the adults, pure, clear colors with a lot of white in them—turquoise, lavender, amethyst, gold, peach, pink, and aqua blue. They are just like fairies, full of Light.

C: It is a similar thing to now, the parents are describing systems and their understanding of this place. They are teaching them how to interact and interrelate with physical density. It's not the same density as we have now. It moves, reacts, and responds to us much more fluidly. We are much more in tune with our surroundings, and we teach our children how that works. We teach them to interact with the plant kingdoms. If we want to grow something like a ship or a dwelling, we show them how to first connect with the consciousness of the new structure or vessel.

CH: Everything has a consciousness that we interact with in a very conscious manner. When we want to create something we enter into an agreement with the consciousness of that creation, whether it is the material that is going to become a building or the entity that is projecting the design. It is a mutual agreement, and everything we do is done by consent and agreement.

C: The buildings, the transport, and everything have the same love frequency, like the way we interact with the time ship. We talk to her telepathically and tell her where we want to go, and she takes us there. That is how we relate with everything in our surroundings.

CH: I am seeing flying cars, and they are using some kind of antigravity device. They are activated by our conscious projection of mental energy; we visualize the place where we want to go, and somehow the car interacts with the electromagnetic energy in our bodies and converts it into forward movement.

CH: I am asking how things are organized for decision making and government.

C: I was just asking the same thing, and the word that came to mind was "consensus," with much less "I am right and you are wrong" approach to government and more "how would you like this to happen?"

CH: I am also seeing councils of elders who are advisors. They don't make decisions; they give advice and counsel. Then the group or people make the decision amongst themselves about things that concern them in their own area whether that is a living area or a work area. Everybody who is involved in a particular process or sphere is also involved in the decision-making process, and it is by consensus. They go to the wise ones to ask for help to make sure that they consider every angle. Then they take that information and make their decision by consensus.

C: When something needs to be done, there is a spontaneous natural occurrence where people gather together, almost as if it entered into everyone's consciousness simultaneously, that now is the time do something and we need to meet. In our time now we are doing this unconsciously where we find that synchronicity brings us into connection with people without making a plan formally. In the future, after the shift, we will be doing this all the time. We just seem to know that it's time to go and meet so and so and we will do such and such.

CH: There is much more telepathy and following intuition.

C: We don't need calendars and appointments anymore. It's more just an inner knowing.

CH: It's like things are beginning to happen now more by synchronicity and Divine appointment.

CH: Let us look at how we grow our food. I am asking if we are still eating food or whether we are doing something different?

C: It seems like we can eat if we want to.

CH: I am seeing very highly concentrated super foods like green algae, spirulina, food from the ocean, green plants, and fruit. Not so much grain and no animals.

C: What has become of the animals?

CH: We talk to them now, they are our friends. A lion just walked in when you said that, we look into each other's eyes and have a conversation. There is so much love and respect for each other; he doesn't think he is king of the jungle, and we don't think we are king of the castle. We are greeting each other as equal Beings in the One Light and under the sun.

CH: Let us look at the contact between the people of Earth and other civilizations. I am seeing very big towers that look like communication stations. Intergalactic stations—very tall—with different levels that ships can come and go from. We can travel out and they can travel in, it is all done with Light ships. There is no rocket fuel; that seems really primitive now. There is a lot of exchange going on; we are helping other civilizations, and they are helping us. It's all about new technology that we are receiving and giving. We are sharing new technology.

C: Yes, there is much more to-ing and fro-ing from different star systems, and it is perfectly normal for us to be doing that. Nobody is worried by that anymore; it's just everyday activity.

CH: There is no fear because we know these Beings from other star systems. We know that they are loving and wise because of their energy. The other energy of ET interference is not on the planet anymore. The cold, mental, scientific entities, which were using humanity and animals for experimentation or to enhance their gene pools, are not on the planet now. That energy was of low vibration and is not sustainable on the Earth anymore.

There is a lot of singing going on. I can hear singing all around the planet. People do more singing than talking; they sing together like celestial choirs. They sing to the plants, the oceans, the animals, and the stars.

C: I was getting the feeling of how birds sing. It feels lovely.

CH: Yes, they just sing for joy.

C: That's a very joyful feeling. I feel as if my heart is flowering. My eyes are tearing up with the idea that we have done it.

CH: I feel like I want to sing this energy back through the dimensional doorway to the time we are in now, and sing it to the people here. So many people are having a really hard time now. There is so much illness, worry, and hopelessness. I want to sing them through the gateway and into the Light.

CH: Let us look at the gateway. Let's track back to the time when the New Earth began.

C: I sense that we are standing in front of it now. Some people pass through from time to time like we have just done.

CH: I see it as a golden archway, but not McDonald's!

C: We go backwards and forwards. There are a number of us who are traveling through it, and we seem to be the road builders or the early pioneers when they were laying the first railway tracks. What we are doing is just opening it up, and the Star Beings are helping us. There is Light pouring in. All of the sunbursts and solar flares are doing what they are doing, but we are here grounding it and making it possible. In the future this kind of travel will be common, but we are laying the pathways and connections in the work we are doing now.

CH: Let us look at the energy that came to the Earth on December 21, 2012, when we aligned with the galactic center.

C: It feels like a bright, sparkly, crackly energy—like fireworks. They are activating sparks; when the light explodes, it feels as if the explosion is inside us.

CH: The analogy that I am getting is that the galaxy is like a spinning top that children play with. We are the point of the top that touches the ground, and all of the energy is being focused into the point after it comes through the portal from the galactic center. It comes from the Great Central Sun, and it is all being focused here on the Earth. Anything that has resistance, solidity, density, and heavy energy is just being blown up as if a laser beam were hitting it. Structures that are more fluid, receptive, and open are opening like a fire flower.

C: I see that you and I are open at this point, wide open. The work we had been doing up until then made us receivers and transmitters of the energy. It flows through us very easily at this point.

CH: I am seeing that there are areas, centers of government, of banking, of giant corporations where there are hubs of energy that are all about control, deceit, and corruption. Those are like black blobs of heavy energy. They are being obliterated. Then there are all these other centers that are Light spheres, such as the children and the animals that are little spheres of Light. They receive a super charge so that they expand and get bigger while the dense, dark ones shatter—it is their own resistance that is causing this. There are points of Light all over the Earth, and those are the people who are awakening or who are willing to awaken. They are bringing the Light in and putting it into the crystal grids, into the waters, into the oceans, and into the Earth. Then it travels around the Earth so everything on Earth receives it. When resistant entities receive it, they implode, and when receptive entities receive it, they expand and blossom. So some people are going, "Oh my God, oh no, what's happening? This is awful," and others are going, "Oh wow! I have been waiting for this all my lives!"

One of the transformations happening is that people who receive this energy become very aware of Mother

Earth as a conscious Being. They want to do everything they can to bring her back into balance, harmony, and peace. Just the idea of doing anything invasive or toxic is now so abhorrent; it's like doing it to your own mother. We are asking what we can do to help. We are starting to clean up the mess we have made. We are experimenting with our new consciousness. We are discovering our powers as souls and are using them to clean the air, the water, and the Earth.

We are talking to the plants, asking how we can help them: "What do you need?" We are also talking to the animals, asking how we can help them and what do they need. The children take to this very easily and very naturally; it's as if they already know how to do this, how to talk to animals, to the trees, to fairies, to plant spirits and elementals. The others are practicing and moving into it, except for people like me who do it all the time! Not many people are doing this now in our time, but it would be good to start. All living things have consciousness and are waiting for us to talk to them.

Now I am feeling this deep sense of peace, like a cloak of peace, a soft golden Light that has just settled over the Earth. It's as if everyone let out one big sigh of relief and release because the stress has just built up and up. After we go through the gateway, it's like, "Aaah, we don't have to carry that anymore."

It feels like we have all been holding our breath for so long, and we don't need to do that anymore. We are not worried about the influence of the people who are resisting the change now, who are trying to hold onto the old ways. There is nothing that they can do to stop it now; it's real, it's happening, it's who We Are. There is nothing anybody can do to change any of it. The only choice they can make is whether they want to make this shift or not. There is no way that they can influence those of us who have already passed through the gateway.

So now the people who have made the shift become like ambassadors. We are showing the new way of dealing with every situation—with love, understanding, compassion, tolerance, and kindness. This is who we are now, and nobody can hurt us anymore. We know that we are now the demonstrators showing the new way. We become role models because we are operating from our hearts, and people need to experience receiving that. We don't need to lecture them or preach at them; we just need to let them feel the Love, and they will want to come with us and do it too. They will say, "I want to do that. I want to live in that place of Love." They can make that shift just by choosing to do it.

CH: I am seeing a lot of activity on the surface of the Earth and in the ocean currents, some wild stuff going on.

C: It seems we are able to observe it and to send it calming energy. There are enough of us on the planet now to be able to calm it. I was reading about stories of devastation, but that we are also the creators so we can decide, "Is this what we want? Maybe it isn't, maybe we would like a nice smooth transition." I can see a lot of us consciously choosing to help the transition at this point.

CH: Yes, I have said many times that we are the midwives and we can assist Mother Earth to give birth gently to her new form by working with her consciously.

C: There is a feeling of that now.

CH: I can feel very strongly this golden cloak of peace that came down. We can focus it and direct it to calm the waters and the fires, to soften the wind and calm the Earth. This is who We Are now. We are this golden cloak of peace, and we have wrapped it around ourselves.

C: We have become it, haven't we?

CH: Yes, and nothing can shake that, nothing can turn it back, it's done now. So the only thing for us to do now is to help others to come to this place and to clean up the mess we have made on the Earth.

C: It's a lot quicker than we think. It seems like it's a hard job and it's going to take a long time but actually when we get into this place and we start co-creating together we realize that we can do it very quickly. A lot of technology has come in from our Star Brothers to assist us that we are accepting.

CH: We can also work directly with the elementals like we used to in Druid times. In fact, the human family has returned to its place in the circle of all kingdoms. We were the only ones who forgot where we were in the circle of life. We can work with the elements and ask them to calm down.

C: I just heard, "Ready when you are."

CH: That's right. They all know who they are, and we are the only ones who have forgotten. They are waiting to welcome us into the circle again. We are all coming Home, and it's a done deal. It's not optional for the planet; she has already made her choice. It is only optional for each individual to make his or her own choice.

C: It's like, "How would you like to experience this? You might as well enjoy it."

CH: Is there anything that people need to know to prepare themselves and their loved ones for this Dimensional Shift?

C: First, start being peace; it begins with one and becomes the many. I have been feeling that more and more.

CH: Drink lots of water, eat lots of green things, and tell people not to be afraid. It's all part of the plan and we planned it well; we covered all contingencies and nothing can stop it. So just pray, ask for help. There are so many Divine Beings on the other planes queuing up waiting to help us. We just need to ask. Mother Earth loves us. She doesn't want to hurt us. She just needs us to wake up and stop doing what we are doing because it is hurting her.

I remember when I was on the Big Island of Hawaii in 2006, and there was an earthquake that shook me out of bed early in the morning. I had no fear because I absolutely knew that Mother Earth was not trying to hurt anyone or me. Only one person was injured on the whole island when a wall collapsed on her arm and it was a 5.6 earthquake. All the neighbors were out making sure everyone was OK, including neighbors across the street that normally never spoke to each other. Some neighbors came by to see if I was OK, and another neighbor had lost her water supply so she came to use my shower. This is the purpose of these events; to teach us to take care of each other and to break down barriers that we think divide us.

Each one who comes Home is the lost sheep or Prodigal Son, and there is great rejoicing in the heavens.

Chapter 6

The New Children Who Came to Help Birth the New Earth

Kathy is an old soul who is about to give birth to one of the new children who are coming to assist those in the transformation of the Earth. She is a powerful healer and teacher. In this session we traveled to some of her existences as a soul where she spent time in other star systems as a preparation for this lifetime upon the Earth. In one of the stars of Orion's belt she connects with the soul of her unborn child and receives information to assist him with his entry onto the Earth plane. She already has a three-year-old son who is of very high vibration.

These new children are masters from a very high dimensional frequency; they need to be supported by having a natural, soft, and drug-free birth and to be surrounded by a nurturing and loving environment. They are already holding the frequency of pure love from the fifth dimension that the Earth is moving into. They are very sensitive to chemicals and to disharmony. They need to be educated in an environment where they are supported to develop their own natural abilities rather than to be forced into a mold that was not designed for fully conscious beings. During this time of transition, as we move from the third-dimensional energy of separation to the fifth-dimensional energy of Oneness, these children need to be protected and nurtured. They are masters of Love, crystalline children who have volunteered to come to Earth at this time to assist in the Ascension.

Kathy also has spent many lifetimes training and working in great temples of healing on the Earth plane in preparation for this time of transformation on the planet. I took Kathy into connection with her Higher Self, who is called Athena, and she was willing to give us information about this lifetime for Kathy.

Charmian—Thank you for agreeing to speak to us. Why did you choose this time in particular to connect with K?

Kathy—It is a pivotal time of transformation in this young life. There are some adjustments that need to be made and some awareness that needs to be enhanced. I come to assist with bringing a new person onto the planet and keeping a balance amongst the household that exists already to optimize the well-being of all those individuals.

CH: Can you give K any information about other lifetimes that may have helped her to prepare for this one?

K: Yes, in Egypt.

CH: Let us go to ancient Egypt and look at the place where K was embodied.

K: It is fertile ground, not desert, there are many Beings there but many of them are not from this plane. There are Beings who are clearly not human and K herself appears to be some sort of combination.

CH: Can you describe K's purpose at that time?

K: She is a clear channel of communication from the stars to this planet, one who is put on Earth to enhance the connection from this plane to the Brethren, the Galactic Brotherhood, clearly not for selfish reasons but to serve the planet.

CH: How did she communicate at that time?

K: She connected with those who hold the vision of evolution for the people of Earth and helped to download and transmit the sacred geometries for the coming years, including the flower of life and the star tetrahedron. She helped by establishing ritual, not for the purposes of religion, but for the purpose of purification of the body temple to step into the higher planes of Light.

CH: Who were the people who were guiding this transformation? Was it a specific body or group?

K: The Pleadians were instrumental, and there was some interaction with the Beings in Orion's belt.

CH: What about the Sirians?

K: K was not connected with them but they did have a presence.

CH: What can K do to bring that energy forward into today?

K: There is a power and a determination that was present in her Being at that point that needs to be re-established now, as well as a connection to the memory of the Brethren. One important way of doing this is to sit out in nature under the stars doing meditation, remembering the deep connection and that the Love is always present and sustaining. She will receive images once more from the Brethren because that will give her a sense of greater purpose and service that is not from her ego. The messages, once heard and felt, sensed and perceived, will be more of a guiding principle than her own sense of what to do.

CH: Can we ask that the Essence from that lifetime in Egypt be brought forward now to activate the dormant codes in the DNA and to reactivate all of her skills and abilities?

K: Yes, we can do that. (We allow some time for the activation to be completed.)

CH: Is there any other message for K at this time?

K: Simply trust, everything is happening at the right time.

CH: Are any of the Beings who were with K in Egypt in her life on Earth now?

K: Many.

CH: What is their purpose together?

K: Remembrance, not because remembrance of the past serves in itself, but simply remembrance of our purpose and our support for each other because not all are in favor of this transformation.

CH: When you meet people who share the memories, will you be activating each other?

K: Yes.

CH: What was my role in that lifetime?

K: You were a guide, not the primary guide but a secondary guide and an example.

CH: Imagine you are sitting out in nature now under the night sky looking at Orion's belt. One of the stars is calling to you. Which one?

K: The one on the right.

CH: You are traveling up a beam of light into the star. It is pulling you into the heart of the star.

K: It's full of light and color, it is so bright and pure. It is elaborate and there is lots of crystal.

CH: Are there Beings in the temple?

K: There is a sense that there would be, but I don't see them right now.

CH: Feel the crystal light activating your crystalline light body. (Pause)

CH: In terms of K's Essence and origins, what role does this star play?

K: It's like an incubation area. It's not her birthing place and it's not her final destination; it is a place for reflection and nurturance. It is an intermediary place.

CH: What does K need to receive here today?

K: Activation of the Light Body—especially at the back of the heart chakra—and purification.

CH: OK, let's see that happening right now. (Pause)

K: The baby has a strong resonance with this place.

CH: OK, let's ask for the child to receive whatever activations he needs to receive here today. What is his connection with this star?

K: This place is more familiar to him than Earth.

CH: Let us ask the Beings who serve in this temple to communicate with us. What message do they have for K at this time?

K: You are welcome here anytime, simply think of this place and you are here. You have spent many lifetimes visiting here, and each time you come you have higher frequencies activated. If you choose, you may visualize this temple around you and know that it is there. It is not simply an imaginary experience. It will help to shield you from lower densities, to increase the amount of Light you draw in and to sustain it during and after the meditation.

CH: What form do you take when you are in this temple?

K: I am very small, almost crystalline and light, not much density to it. It is not like air. There is a form that has substance and a presence.

CH: Is there a crystal on Earth now that would help K to keep this connection?

K: There is a crystal in South America in one of the Mayan temples but that will not be available. The stone which does not have the same frequency, but which would be of assistance is labradorite.

CH: Let us ask to be shown the form your baby takes in this star.

K: He is almost like an ascended master; he is more ephemeral and larger.

CH: What is your relationship with him here?

K: He is my guide.

CH: What is your agreement about him being embodied on the Earth plane?

K: One in which I bring him into the world and nourish him and allow him to fully embody who he is. He is a powerful soul. We need to choose an educational system that will allow him to make and support his own choices. We need to support him in his growth without him becoming a sheep like so many other people on the planet. He has a clear voice; he has a knowing and that knowing will guide him more than any direction ever would. It is important at this time to honor one's being as a mother and also to know that a mother can't fully know her child unless she had first known him as herself.

CH: Is this why K was brought to this star today, to connect with the soul of the baby?

K: Yes.

CH: What does he need for his birth experience?

K: He needs music and energy to be sent ahead of time to the place where he is to be born. There is a lot of fear energy in that place which needs to be cleared. Having salt in the room would be good for clearing as well as proper lighting that is slightly pinkish. It is important to intend that those who are aligned with the highest and best good will the birthing attendants and to clearly intend that only those who are calm, present, and centered enter the space. He will birth with little ado, so be ready. It is important to mentally prepare and relax. Imbibe warm drinks throughout the birth and now. Miso soup would be good both at and after the birth.

CH: Let us ask now for a bowl of perfect Love to be created in the room where baby will be born; the perfect birthing and receiving bowl for this Divine child to birth into in a most gracious, easy, soft, and gentle way. We see this in the name of the One and give thanks that it is so. Feel your two energies in the star merging and coming into complete harmony.

CH: I teach a program called Hypnobabies that teaches mothers to stay relaxed during birth by using hypnosis and to visualize the birth they want. Would this be of benefit for this birth?

K: Definitely.

CH: Does baby want to say anything?

K: Read the book, *Initiation*, by Elizabeth Haich, avoid media, spend time outdoors in nature, nurture the downward movement by squatting, and visualize the birth you want. Remember that any and all trials at this point will pass. Do not place much value in the way things look or seem. They are all lessons designed to fade. There is not much time between now and the birth. People call it labor because it is hard, but it is simply a process. When no fear is present there is no hard work as the body knows how to do this. I need not fear for my mind or my body. I have my path and All Is Well. Do not let anything cause tension. I am whole and complete.

CH: I see his soul is of a very high frequency, and he is orchestrating the plan for his birth.

K: He says, "Do not fear for my brother, he and I are great friends. We will uplift and inform each other. Yes, there

48

will be some small challenge from the ego simply because he has had all the attention but overall he and I have great love for each other and all will be well."

CH: Is there anything else he needs to receive from this star?

K: No.

CH: Is there a way for K to empower her new frequency?

K: The practice of becoming the Goddess Durga.

CH: Is there memory from Hawaii that will help to download this new frequency further?

K: I have a vision of a gathering of Beings around the great Light. There is a smell almost like flowers but different, a peacefulness. It is the memory of the Rainbow People.

CH: Who are they?

K: Before they were in physical bodies they were the Rainbow People, sort of a bridge from Lemuria.

CH: Ask to be shown the form that K had as a Rainbow person.

K: It's almost not a physical form, as if I myself were a guide.

CH: What were the Beings doing around the Light?

K: They were anchoring the Light on the planet and connecting it to different power centers.

CH: They were the grid masters; they brought in the light codes. They were already in Oneness and then they came down into density until it was going to be time for the cycle to be complete, for the Earth to return to the Light.

CH: Tell me how the Lemurian memories are important for K and for the planet?

K: It is important because of the vibration of the shift. From whence you came to there you are going once again. The connection to the deep memory of that vibration, that is the deeper essence than the physical body, helps to clarify the vibration of the physical body and the vibration of one's energy centers. Remembering this helps you to connect to the people who are here at this time. There are many on the islands.

CH: What is the significance of Hawaii in the Lemurian memories?

K: Hawaii is the Place, and there are many people who are attracted to that. Those who need the frequency are attracted to come here, whether it be for a short time or to be here indefinitely.

CH: Let's ask about the end of the cycle because this is why we came. The whole experiment was for us to make this Dimensional Shift. Let us ask to be taken forward in time to when the Dimensional Shift is happening on the Earth.

K: Before this can happen, things that no longer serve, things that don't fit with that must rise to the surface to be eliminated. This is why people experience so much transition in their lives. It is important to remember that it is simply a matter of events rising that don't coalesce with the higher vibration that must be purified and released. It is easier to do this in meditation and harder to do it through physical experience. As the vibration is rising, with more and more people choosing to consciously transform rather than unconsciously transform, the planetary tensions lighten up. The connection of humans to nature becomes deeper. Humans begin to remember their place in the system rather than trying to create a place in the system. There is a settling of the nervous system, a little bit counterintuitive because one would think the nervous system is speeding up. Actually as the nervous system reaches these vibrations that are more its birthright, there is a settling, a calming of the organism. This calming creates more interconnected interactivity, also more synchronicity, more harmony so to speak in interpersonal relationships.

CH: As we look at, December 21st 2012, what was the significance of that time in the rising of the frequencies on the Earth?

K: It was a pivotal point. It was not The Point of transformation, but it was a pivotal point at which there is a higher rate of positivity than negativity on this planet. That does not mean just amongst humans, but there is more positivity that is not even associated with people. There is a greater rhythm that feeds the acceleration from 2012 onwards.

CH: After this activation, how are people different to the way they are now?

K: People are more aware, and there is a yearning among people who perhaps didn't have a spiritual inclination before. There is a yearning to connect with something deeper. Many left during that year of transition. Those who were not ready to access this higher frequency or openness chose to leave and to go elsewhere. That's fine; that is a choice they have made. There are many places to go in this universe and others, where they can continue their transformation at a greater density, but not on this planet. This planet's time is now. She is in need of great Love and, at the same time, is full of this great Love.

CH: Is there anything we can do to prepare for this shift?

K: Our own personal practices—that is practices that we do out of Love and out of nurturance of the planet and ourselves. Listening, as well, is important; listening more and more, not only with our ears but with all the senses. Also noticing synchronicity is important, being aware of when things are falling into place. When there may be the slightest inclination that something is slightly off, really listening to those internal signals. That is when a greater shift can take place before anything major occurs. So one of the many things that people need to do right now is really connect to their own nature and desires rather than what they may have been told by their family, by society, by the methods which they grew up with. It's really essential to connect with one's own soul destiny instead of the messages that come from the outside. We need to realize that, when one is connecting with the intergalactic family, the messages arise from the inside even though they could be thought of as coming from the outside.

CH: You said December 21, 2012, was not the final event. Is there a time when the Earth has completely shifted and everything on Earth is in the new frequency?

K: This will take some time and remember that frequencies are always shifting so there is no point of finality. There is always an evolutionary process of expansion in the universe from here to there or there to there. It's like entering kindergarten then first grade and second grade;

there is always somewhere more to go. Always somewhere more beautiful and more desirable, even though it is important to savor each point as one is there. Savor and appreciate this time of transformation. It is beautiful on this planet.

CH: As the Earth is rising into the new frequency of Ascension into a higher dimension, is there a time when people who are not in alignment with the higher frequency will be leaving?

K: Yes, it is impossible to survive in a climate that doesn't suit one, it's impossible to survive in an atmospheric condition that one was not designed for.

CH: How is all of the financial unrest and political upheaval in the world serving the Divine plan?

K: It is serving in a beautiful way. It is helping people to remember that power is not from the outside, that regardless of whether a person has one million or one billion, it doesn't make one happier. It is helping people to remember that all these things are, in fact, illusions. The illusion that you invest some money and that it grows because of whatever circumstances—it is all illusion and money itself is simply energy, a fact that people have forgotten. True power comes from our connections—connection to nature, connection to others, and most importantly connection to the Higher Self.

CH: Let us look to the time after 2012, after the shift. How are people living on the Earth then?

K: The structures that are in place, the financial and the political systems, have taken some time to be created and they will take some time also to be eliminated but they ultimately will be. Instead, people have a greater reliance on their own self-worth—worth from within—instead of relying on someone to provide wages or health insurance. Health insurance comes from connecting with nature and what one's body needs. That is true insurance against illness. Speaking of health, humans—that is, people coming from the Earth—will remember that it is important to have health emotionally, spiritually, and mentally first, before physically.

CH: As the consciousness of the people on Earth is lifting up what will happen to the illness that people have now?

K: We will be able to maintain perfect health.

CH: In Lemuria we worked in light teams to manifest every-thing we needed directly from the Source by receiving Creation Codes and using our intention and visualiza-tion. Do you see this gift being reborn?

K: It must be reborn. It will not be a process that we will see in twenty years, but it will be reborn. It is an innate part of the Beings that we are. Even birthing has become of such a great density that it causes people pain, which was never the intention. Birthing is an act of great Joy. Birthing of anything—not simply a human baby but an-ything that we create in any positive way, ushering in a soul coming in to experience this planet for the first time—is one of the greatest gifts, and yet people associ-ate it with great pain. What a tragedy!

CH: Is there anything else that Athena, K's Higher Self, would like to communicate to K at this time?

K: What is important at this time we are in right now is the dropping of fear—fear about any situation, any commu-nication—knowing that All is well. All is well every mo-ment, with every experience. There is nothing in your life worth fretting about for you are always loved, guided, and supported. Trust your inner wisdom that the guidance that has been there for generations, for lifetimes, for lives on other star systems, will continue to support us at all moments, everywhere.

Chapter 7

Beings from Many Dimensions Now Embodied on Earth

Many Lightworkers here on the planet are from the Angelic planes. They have come from Oneness and Light into darkness and density to assist Gaia and humanity to make the Dimensional Shift from the third dimension of separation to the fifth of unity. Britta is one such Being. We connected to her Higher Self. When I requested that she be taken to a lifetime that was a significant preparation for this present time on the Earth, we were connected with the Angelic planes and to Venus. Many Lightworkers had experience of absorbing the vibration of pure Love on Venus that they then brought to the Earth.

Britta has two boys, both very high souls, who have volunteered to come to the Earth at this pivotal time to assist with the transition from darkness into Light. Britta looks like an Angel. She has blond curly hair, bright blue eyes, and a beautiful classical face. Her gift from the Light comes to Earth in her work as an artist and also as a jeweler, creating powerful jewelry with semi-precious stones and beads, which help to open the chakras and balance the energy in the body.

Here is the story of her session: I asked her Higher Self to show her an existence that was a preparation for this time of renewal on the Earth and to help her to remember who she is.

Charmian—Why have you chosen this time to connect with Britta?

Britta—She is stuck and needs some guidance.

CH: Can you take Britta to a past life that will help her to remember who she is?

See yourself floating on a cloud over many landscapes, different lives and different times in different civilizations. You are coming down in a lifetime that you need to remember. See the cloud coming down in another landscape and describe what you see.

B: I see lots of buildings like a fantasy. They look like crystals, big crystals with golden tops; golden roofs and golden symbols on the top of the buildings. There are a lot of these buildings everywhere. They almost feel like glass. The crystals are transparent and gold.

CH: Look down at yourself, what does that look like?

B: I still see myself as I look now—blond curly hair—but I also see the Light that I am, a golden, clear Light.

CH: What are you wearing?

B: A white goddess dress.

CH: Let us go to the place in this crystal city where you do some important activity.

B: I am going into one of the buildings. I am wearing this robe that is on one shoulder. It is asymmetrical and I have something gold around my head. I am walking to this temple. It's beautiful inside. I feel peaceful; I feel good; I feel purity, quietness, and wisdom. I see statues like quartz—like geodes with crystals inside—there are lots of those around me too.

CH: What have you come to do in this temple? Is there a particular alignment you are waiting for?

B: My Light is in the temple, but the temple is Light as well.

CH: Where do you feel you want to stand?

B: In the center underneath the triangle at the top of the roof. It looks like a pyramid as I look up. There is a hole in the apex of the triangle.

CH: Let us ask to be taken to the moment of the alignment when a beam of Light comes in through the hole.

B: It's like a star; I am waiting for a star to pass overhead. The sky is darker now. It is nighttime. I see a big, bright star shining through the hole, shining down on me. I see other stars, but I am drawn to this big bright one. That is my main star, and I am waiting to connect with it. I am looking up at it through the hole.

CH: As you take the beam into your body, feel the connection through the top of your head and bring the energy down into your whole body. Do you feel you want to go up the beam into the star?

B: Yes.

CH: Let the beam of light draw you up into the star. Do you know which star this is?

B: It is Venus.

CH: Allow yourself to float into the dimensional temple in the heart of the star. The beam of Light brings you directly into this Venusian temple. What do you see?

B: It is a diamond shape in the middle of the star. I am going inside the diamond, right into the middle of it.

CH: How does it feel inside this star?

B: It feels warm and light, a strong light; the light is gold with sparkles of gold floating in it like golden snowflakes. It is dreamy and pretty and magical. I can make my wishes come true.

CH: Why has her Higher Self brought Britta here today to this Venusian temple?

B: To show her a Higher Self than herself.

CH: Do you have a family here on Venus that wants to connect with you?

B: I do see two boys; mainly the younger one with white blond hair, sparkling blue eyes, and white skin.

CH: What are his qualities and strengths here on Venus?

B: He is strong, very strong, but also very loving and pure.

CH: Is this your son on the Earth?

B: Yes.

CH: Why has he chosen to come to the Earth at this time?

B: There is purity there.

CH: Look at the form he takes as a Being rather than as a child on Venus. What does that look like?

B: He also has a golden Light, a see-through golden Light. I also see blue, a see-through aqua blue light.

CH: Look into his eyes. What do you see?

B: Lots of love.

CH: What is your relationship here on Venus?

B: Pure love, a lot of love.

CH: Why did he choose you as his Earth mother in this life-time?

B: For us to be together.

CH: What does he need from you as a mother now to help him to bring these Venusian qualities of pure Love to the Earth?

B: I need to open to him, to receive his energy, to let his feelings come out, and for him to explore them at a deeper level.

CH: Is there something in this temple that you need to bring back?

B: A red ruby.

CH: Put the ruby into your heart. What does it represent?

B: There is strength, a purity, fire, creativity, and magic.

CH: Is there something for your son to bring back?

B: It's a blue stone, clear blue like aquamarine or a blue opal.

CH: What are its qualities?

B: There is lightness, strength, and a connection to the ruby. The stone is almost as if it is him. He is that stone; he is that gemstone. It has strength and wisdom but also softness; it is loving and warm.

CH: Come back down the beam now with your son and the stones into the temple where you began. Do you know where this temple is?

B: It is not on the Earth. It feels magical like in a fairytale book. It's dreamy; it's white. It is a crystal building with golden symbols. It feels angelic. It's the Angelic plane; white flowing dresses and curly golden hair. It is light and floaty.

CH: Know that this is who you really are. You are not limited by anything on this Earth plane. You can create whatever you want with your own magic. You can create your world the way you want it to be. Is there anything you want to bring back to Earth with you from this place?

B: Yes, the gold headdress. There is a jewel in the center. I see a green emerald with diamonds around it. It's more like a crown.

CH: So take the crown, it is yours. What does the crown represent?

B: It represents the magical place that this is. There is a presence, a purity, a heavenly feeling, this whiteness, this golden angelic feeling. It is beautiful and it is safe, and this crown represents that beautiful place and its magic.

CH: The crown represents your golden self, your True Being. What quality does the emerald represent?

B: It feels like it has been passed down to me, it has been passed down and passed down. It was given to me at a certain time in my life, and it was placed in the middle of the crown.

CH: Is it a symbol of an office or an achievement?

B: I was taught things and groomed; I had to learn things. There were lots of women, goddesses and angels. I had to go through training. I learned what I needed to learn, and then when I got to the place where I needed to be I was gifted with the emerald crown.

CH: Was it a symbol that you had graduated?

B: It has something to do with powers. I had to be taught, I had it when I was born, as a girl and a grown up. All these women taught me the things I needed to know. I went through a long training but it was nice, very lovely, peaceful, beautiful, and magical.

CH: After you received the crown, what powers did you have? What were you able to do?

B: It looks like touching people with my Light, with my hands, and with my Light.

CH: Why did you choose to bring this gift to the Earth?

B: To help people. To give them purity and the ability to look within themselves to see their own honesty and their own Light. To help them open up to the Light they have within themselves. That Light has to come out and be seen.

CH: At this point in Britta's life, how can she put this gift into practice?

B: By having conversations with people, communication. I have a passion for painting, I feel so strongly about that.

CH: Let us look at the effect the paintings have on people when they look at them.

B: They are bright, colorful, and calm. I paint a lot in aqua blue, turquoise, and white, and I put gold flakes in them.

CH: Those are the colors you saw in your Venusian and Angelic temples, so you are bringing the energy of those temples into your paintings. They will activate people to remember those planes and to remember who they are. What about your jewelry?

B: I hand pick gemstones at the shows. I bead them, but I need to study more about the relation of gemstones to the chakras.

CH: That was part of your training as a priestess, making amulets for healing and initiation, not just adornment. Your artwork and your jewelry are both helping you to bring the energy of the Angelic planes to the Earth.

CH: Let us move forward in time now to the December 2012 alignment of the Earth with the galactic center. What do you see?

B: I see energy all around the Earth. There is a little bit of shaking going on. I do feel a little nervous, but I feel I would have to turn to my Light side, my sword of power, my temple, and my crown. I know that I am protected in my Light, and my children and family are protected as well. I feel that I want all of my loved ones to be protected, that includes my friends and family—whoever is close to me. I want a white Light all around us.

CH: What does this shaking look like?

B: It could be an earthquake. I want to think that we won't be in the place where the earthquake is, but it's a bit scary. That's what is making me nervous.

CH: Ask your Higher Consciousness if there is a way that the shaking can be calmed, a way that we could help the Earth with her transformation without such dramatic physical events?

B: By loving Mother Earth, praying to her, respecting her, loving the water, the trees, and the flowers.

CH: Look at the whole Earth and at all the Lightworkers who are preparing themselves to make this energy shift. When the shift happens, how will the transformation be for them?

B: They will be protected.

CH: How are they helping the shift? What is their role?

B: By talking and sharing with others, they are passing their knowledge on. They are telling people to be even more grateful for the world, for nature, and more thankful for what they have. The Earth needs us to respect and love her.

CH: What about the people who are not focusing on their Light just now? What happens to them?

B: I think they are going to get lost. I see them falling off the Earth, not being connected, being lost, and not getting anywhere.

CH: What is the most important thing that people can do right now to prepare themselves for this Dimensional Shift?

B: Remember to be loving and kind to everybody, to nature, and to people. To be just our pure, honest selves everywhere we go. Be nice to everybody we encounter, be light, peaceful, and respectful. Pass the Light along, pass the torch.

CH: Let us look at the time on the Earth after the shift has taken place. How are people different?

B: People are loving and kind, and they look deeply into each other's eyes—they make a connection. They are being honest and real with each other. They are having beautiful heartfelt conversations, laughing and dancing. They are free, full of light, and happy. People are not so serious; they have more joy. There is peace, love, and kindness. I'm not sure it's exactly what it is like to be in Tibet or India with all the monks who meditate all day, but there is that peacefulness here, like that loving, caring, Buddhist way of living their lives. People have become very disconnected now, and they need to be connected again. I do feel that I am pulling away from people whom I feel no connection with, and I like being with people whom I do feel a connection with. It is important that we spend time with people who will help to lift us up rather than with those who are negative and will bring us down.

CH: What about the children, how are they connected to the New Earth?

B: They are all different colors, all the colors of light; they are like gemstones. Each one has his or her own energies and qualities. I see them dancing around, singing, playing, and skipping. They are dancing around the world putting the magic back into the world. That magic is coming back to the Earth. They are teaching us, and we are learning from them.

CH: What is the difference between their soul vibration and ours?

B: Theirs is stronger. They have a stronger connection, and they are much more powerful.

CH: Their connection to what?

B: To the Source, but also to themselves. They know who they are; the older generations have lost the connection. I have been connected. I may have gone off track here and there, but I always felt—even as a little girl—I felt the connection and a true belief in loving kindness and honesty. It is a feeling of being honest and true—even if the truth hurts sometimes—but not to be superficial.

CH: Now that you have had this reconnection to your Higher Self, how can you pass it on to others to help them to reconnect?

B: By having conversations with people who are open to having all kinds of conversations. Talking about all of these things, about our different views, and somehow there will be a mutual understanding. There will be a familiarity between us, and it will feel right.

CH: Know that your Higher Self will be looking out through your eyes now so that when you look into someone's eyes you will be seeing God in them.

CH: Is there a last message from Britta's Higher Self at this time of transformation on the Earth?

B: Continue to be open and aware, trust all of the feelings and intuitions that you get. Bring those around you back into the Light. Learn more about gemstones and keep doing the paintings from the world that you come from. Keep affirming those around you as being in their Light presence and that will help them to be there.

Chapter 8

A New Earth Rising

Mary Carol Breckenridge is an elderly lady living in Hawaii. She is an old Lemurian soul and has been on the Earth in service to humanity from the beginning. She has very clear access to the Akashic records and was able to bring forth much information about the coming Dimensional Shift. I guided her through the time of the Dimensional Shift as we transition from the old way to the new, and here is the information she brought through about the New Earth. Mary Carol has a very close connection with her children and grandchildren, and she was concerned that they would all be making the Ascension. She saw the New Earth being born and the Old Earth separating until gradually there was no connection between them. Mary Carol was able to see the New Earth arising out of chaos and people living in harmony and peace with each other.

Charmian—Mother Earth called the Lightworkers. How did you respond?

Mary Carol—I offered to be of service to the Light and to bring transformation.

CH: You agreed to come to the Earth?

MC: Yes.

CH: Did you and I have an agreement to come together in 2012?

MC: Yes, we made a date.

CH: Let us look at the Dimensional Shift of December 21, 2012, and the changes that are activated on the Earth. What does that look like?

MC: You come through together. It looks like ascending, elevating, moving into another environment of Love. There are other lovely people here too. It's like going through a tunnel, ascending through a tunnel of energy that is embracing us and yet moving us upward. It's a different level of being where it feels good.

CH: What happens to the body through the tunnel?

MC: It turns to Light—two lights embracing each other.

CH: Are you still in a physical body?

MC: Not at that moment. We are moving, and we are transformed into Light as we move. We are moving through another energy field, and then we are going to have a new body that is similar to the old one.

CH: Is the material the same?

MC: Not quite.

CH: Do you look like you look now?

MC: Yes.

CH: What happens to the parts of the body that are painful now?

MC: Everything is perfect.

CH: Let us go out the other side of the tunnel. How are people living on the Earth?

MC: Very differently—there is love, peace, and harmony. Everything is provided and everyone is helping. There is still some work we are doing to heal Mother Earth. We are doing our work.

CH: How are you doing that?

MC: We love Mother Earth and all of her creations. We are cleansing everything with Light, cleaning the oceans, the air, everything. We are making everything clean, pristine, livable, and renewed.

CH: In these new bodies, how do we sustain ourselves?

MC: Very lightly—fruit, water, and breathing the wonderful air and being with the energies—understanding the nature spirits and being in harmony with all that we absorb energetically.

CH: Do you have a particular purpose with any of the kingdoms?

MC: I work with the Devas, the plant people, the tree people, and the fairies who live in all the flowers. I still play in the ocean with the dolphins.

CH: Do you communicate with the dolphins?

MC: Oh, yes.

CH: What are they telling you?

MC: Thank you for returning. Thank you for cleaning the ocean. Thank you for keeping us here.

CH: Imagine you are flying over the Earth now looking down. How is it different to the way it looks now?

MC: She is more radiant, she sparkles, she has no dark clouds, she has whatever atmospheric protection is needed for everything upon her surface to thrive.

CH: Is there pollution, factories, chimneys?

MC: No, there is no more pollution. There is only serving the Light, serving one another, and intergalactic connections. It is easier to teleport, so there is more visiting with our brothers and sisters far away.

CH: What kind of structures are people living in?

MC: The whole Earth is one community. The structures are earthen; they are made from natural materials like bamboo and clay. They are very individual, and when anything is taken from the Earth to make dwellings, an offering is made and is gratefully accepted.

CH: How many people made this Ascension? Did everybody go?

MC: About half of the population, maybe less.

CH: What happened to the ones who didn't go?

MC: They went somewhere else.

CH: What happened to the Earth that they were living on with all the factories, gasoline, cars, smoke, and dirt?

MC: That energetic body has gone to different parts of the galaxy.

CH: So the low-vibrational energy has been taken somewhere else?

MC: Yes, to another station, an area for clearing, cleansing, or even dissolving.

CH: What happened to the people who did not ascend?

MC: Their souls went back for revision.

CH: What is the connection between Lemuria and the New Earth?

MC: The consciousness, the energetics, and the revelation of how many of us came to be. It is the revelation of how things can be in peace and harmony, and it is helping us to raise our consciousness into the fifth dimension.

CH: When we remember our own experience in Lemuria, how is that memory helping us now?

MC: It helps us because we know how to live in Love and harmony. We know how to create; we know that all of our needs can easily be met; we know that we are here to help others; we know that we are joining with others to form a heart sphere of Light.

CH: What is the significance of Hawaii?

MC: It is a Light Center of the Divine Oneness. It is where Lemuria was. That civilization infiltrated the land and the waters here, so the energies are still here from that time. The Hawaiian and Pacific people retained much of the knowledge in its original form. They retained the understanding of how to live gracefully on Mother Earth, and now that energy is here.

CH: So walking on the land or swimming in the ocean is helping us to remember?

MC: Absolutely.

CH: What role do the dolphins and whales play?

MC: They hold the energy, they keep it here. They have activated Light bodies at this time, although we can't see the Light bodies.

CH: Do they see our Light bodies?

MC: Yes. They did not lose the memory of their divinity like we did, so they are helping us to remember the Light.

CH: How are we to work with physical issues before Ascension?

MC: Acknowledge them and know that there is a reason for them and that they are being cleared. It may seem slow, but you will have more clarity on every level of your being. Trust and be sure it is happening now, even though it may seem slow, but it is so.

CH: What happens between the Old Earth and the New Earth?

MC: There is a parting—the New Earth stays in our galaxy and the Old Earth goes to another place.

CH: Do you mean there is the separation of the two Earths?

MC: I just saw it, there is a physical separation. This is the New Earth, and the Old Earth goes somewhere else to regenerate.

CH: Where was Charmian on December 21, 2012, when the shift happened?

MC: She was in Mexico. She was transported somehow, but she was here and there. She is capable of being in more than one place at a time. She was in several places. She has been here on Maui and her spirit is here; she has done her work here and there is more to do. She goes to places where she needs to be; she is called and she goes. She takes the Lemurian frequency out wherever she goes because she is a grid master.

CH: Where will MC be on the New Earth?

MC: I see a beautiful garden; I am wearing silk; there is a beautiful temple on a bluff overlooking the ocean. I am wearing loose-flowing robes.

CH: How old are you?

MC: I am ageless.

CH: What do you do in the temple?

MC: It's a Light dance.

CH: As you perform the dance, how do you feel?

MC: I feel like I am sharing the Light.

CH: What effect does the dance have on the people who are watching it?

MC: It helps them to shine in their own way.

CH: Go to an important day in this time.

MC: We have the visitor, a Supreme Being. He takes on a physical form and becomes my twin flame.

CH: Can you describe him?

MC: He is still taking form, a reformation from the descending; it is the condensation—that's what it is—just a Light form molding itself, condensing into the form of a very beautiful man. He has brown hair, not too tall, he is slender, very graceful, and very stately at the same time. He is very happy. He smiles a lot. He is filled with joy. He is also wearing a silk robe, but his is tied at the waist. He sings and plays the flute.

CH: What do you do together when he comes?

MC: He sings and plays the flute while I dance.

CH: How does that feel?

MC: It feels sublime, almost like swimming. There is a great pool and the dolphins come into it; they love to hear the flute music, and they sing.

CH: What time on the Earth is this happening?

MC: It's in another dimension.

CH: Is this connected to the Dimensional Shift? Let us ask to be shown the way through the process from now until the Dimensional Shift.

MC: The Great White Brotherhood respond: "You don't need to know the way because the way knows you."

MC: Do I need to know or do anything more? Can I take my children with me?

GWB: "It will be their choice, but they will come with you."

CH: What do we need to let go of in order to prepare for the Dimensional Shift?

MC: The energy of fear and inflexibility about needing to make things happen, to be in control.

CH: Is it about surrendering?

MC: Surrender, surrender, surrender to the Light. Surrender to what is, and it's about being light hearted. You are light hearted, so just be what you are and let go of whatever you think you need to hold onto. You don't need to hold on, just let go and it all comes to you in the way that is most perfect. All that is yours will come to you because of your peace, your love, your joy, and your harmony with what is now. The energy after 11.11.11 is all about surrendering, trying to do anything; surrendering letting go of trying and that makes space for the Beings who are helping us to take over and doing things for us. When we are still trying to "do" things, we are blocking access to the Beings who have access to huge amounts of creative energy including our Higher Selves. When we get out of the way and ask for them to do it, surrender everything to them, then miracles happen.

CH: At 11.11.11, I was given an enormous crystal bowl. In the bowl I had to place everything that I wanted to manifest and everything that I was attached to. So imagine now that you are being given this enormous crystal bowl. It is full of the Light, it is sparkling with Light. Put into the bowl everything that you need at this time in your life and surrender it to the Source. When you have placed all of your needs, desires, and attachments into the bowl, the Angels and Beings of Light come. They are lifting the bowl, lifting it up. Everything you have requested and surrendered will be put into manifestation now; it is not your responsibility or concern anymore. All that you have been concerned about in the past—all is forgiven, all is well. We are here with you now.

CH: Mary Carol has some regrets about past choices in her life. Can you help her to let go of her regrets. Why did she leave her children's father?

MC: She had her soul work to do, and she had a different path. It was to take the Light into places where there was no Light. It was all about taking the Light and at the same time bringing the Light to her loved ones, bringing more Light by spreading the Light. She was able to bring more Light to her family. It's all good, it's all okay, and there is no need for regrets.

CH: Leaving that relationship she gave up financial security. How is she to find it now?

MC: It's on order, but it's by not looking, it's by letting go and accepting and letting others help. Help is on the way.

CH: It's about trusting and having faith?

MC: Yes, and letting go of "Oh dear, when is this going to happen? Oh dear, how is this going to happen? Oh dear, what do I need to do?" Let go of all those "oh dear's." Let go because it's all on order.

CH: Why did you show Mary Carol the vision of the future with her twin flame on the fifth dimension? Why did you want her to experience that today?

MC: Because it is very beautiful and it will come to pass and she need not wonder, "Oh dear, will it really happen?" She needs to let go of all that because it's all on order. Part of that was the past time, but it is also a now time that is revealing itself in its own time.

CH: Is it to do with Lemuria coming back? Will we be living the way we lived in Lemuria before The Fall when we lived in Oneness and Light.

MC: Yes, it is all that and more.

CH: Looking at the frequency of the Earth, tell me how the Earth changed between 11.11.11 and the Dimensional Shift? 11.11.11 was a frequency shift. How did that affect the Earth and the plan for the Dimensional Shift?

MC: It has to do with the Light stations, there are Light stations all over the Earth that were activated. Not just on Maui, they are everywhere.

CH: What happened to the Light stations on 11.11.11?

MC: They lit up.

CH: Is my home one of the Light stations?

MC: It was already identified before as a temple or Light station.

CH: After the Light stations are all lit up, how will this affect the planet as it is now?

MC: It's the shifting of the frequencies; it is surreptitiously making a new grid on top of the old grid.

CH: Is there more than one dimension to the new grid?

MC: It is multidimensional. It is very light; it is like a transformer; the Light stations look like big transformers and the whole grid is a transformer. It is very crystalline.

CH: What happened to the old grid?

MC: It has been disintegrating, and it is being replaced.

CH: What was it about the old grid that made it no longer sustainable?

MC: All the energy pollution, hate, war, killing corroded the grid. All that was in darkness. All that was not our soul's highest purpose. All that was not of the Light that was on the planet.

CH: What effect will this have on our planet?

MC: It will change our whole way of living. It will change the way that we are with each other—the way that we adopt the energy of the grid. Peace and harmony will simplify and glorify all that is ours.

CH: What will happen to the institutions that are based on the energies of the old grid, the energy of greed and the abuse of power?

MC: It's like a dissolving, an energetic transformation from an old way of perceiving and doing to the new way of being—living in peace and love and harmony. It's happening now and things are not the way they look. It's more insidious, but in a good way. There is so much Light and so many beings guiding, clearing, and helping our souls to a place of peace, unification, and Oneness. Anything that is not in Oneness is leaving; it is dissolving.

CH: Earlier you said that the Old Earth was going to split away from the New Earth? Is this connected to the process of dissolution?

MC: Yes, that is part of the dissolution so that we won't be disillusioned by the dissolution, but we will be glorified by the dissolution. It's like a separation; it's an energetic event that is hard to describe—a parting. The parting of the ways.

CH: Is that a dimensional parting? That some people will be shifting to another dimension and some will not?

MC: It seems so, but it's their choice. They can all come, but they have to be willing to change the old ways—they have to let them go.

CH: Many people are experiencing fear right now because the financial systems seem to be disintegrating. Can you give them any help as to how to move through these last days before the shift?

MC: Just let go and let God. Let all the Beings of Glory do the work and don't try so hard. So many people are trying too hard to make things happen, and it's all very good, but everything is going to change anyhow. If your heart is in the change then that is helping. So I would say, just surf along with the waves of change.

CH: I did a session with someone recently who saw that there had been a great flood, and a lot of people were washed away, and a lot of places were under water. Do you see anything like that?

MC: I see more of the flooding of Light, the washing, I don't know if it is the rising of the water, it's more like Light rising.

CH: As this Light washes over the planet what happens to the people as they receive it?

MC: They either surf with the flow or they try to hold on to the old and they get washed away. You can't hold on; you have to let go and move with the flow.

CH: After this washing of Light what happens to the Earth?

MC: She is so happy, and so clean. She sings and all the beings sing, the birds sing, the trees all sing. Oh my, is she ever happy!

CH: What about the people on the Earth?

MC: They are all singing and dancing, rejoicing and loving and having a good time.

CH: How are they living?

MC: Oh my goodness, it's a way of living in Peace and Harmony and in tune with Mother Earth and all that she has to offer and give to us. It's the true way, it's really tuning in to our Earth Mother. We are shown how to be, and we have guides in this new creation. All is in place, and all is the same but it is also all new.

CH: Are we still cultivating the land?

MC: Yes, but it doesn't require machines.

CH: What does it require?

MC: Thought forms, dissemination of the thought forms that plant the seeds where they need to be.

CH: Is there any form of money or exchange of some kind?

MC: There is no need for money. There is an exchange in other ways, lots of that is being done already today, people are trading.

CH: So it will be more like bartering and exchanging than buying?

MC: Giving and receiving is our soul exchange, our Godly exchange.

CH: This is how we did it in Lemuria. If we were gardening we saw with our minds what was needed and the soil would move aside if we needed a hole to be dug. Is this why Lemurian memories are coming up now because we already know how to do this? Are the memories helping to prepare people?

MC: Of course. It's the Great Mother Goddess of Lemuria who is blessing us collectively. Those of us who are ready and open to receive, we are awakening to remembrances of how easy is it is to live on our Mother Earth when we are in harmony with her. By tuning in, we are One. The Light towers are like tuning forks, they help us to get tuned in. The people who don't tune in will be tuned out.

CH: So these were the Light towers that were activated on 11.11.11?

MC: Yes.

CH: So, they are now tuning centers to tune people into the new vibration?

MC: Yes.

CH: Are there more frequency changes coming after the Dimensional Shift?

MC: It's a flow—it's not like bing, bing, bing—it's a continuous flow. Sometimes it's high and sometimes it's low. 11.11.11 was like a gateway. It amplified the frequency like a transformer.

CH: Can you give Mary Carol a message from her future self to help her through the times to come?

MC: Keep on knowing that all is in Divine Right Order, all is changing for those willing to change and transform. All is moving forward in a direction that cannot be seen, only imagined. That is all good. Everything is moving fast at this time. It is an energetic flowing—like the Hopi say about the river, "Don't hold on to the shore or try to get out when the river is flowing fast, go with the flow." And, even if it sometimes doesn't seem to be flowing, you just float and you know to let go of any peculiarities, doubts, or fears. Let them flow away, know and keep knowing what you know and doing what you do. Shine your Light and have good faith. Know that your dear family will be reunited.

CH: Do you mean those who have already passed over or families who are still in the body or both?

MC: All of those I know who have passed over in this lifetime, but especially and including those still living. They will all be together. Mary Carol will be with her beloved daughters, grandchildren, friends, and relatives.

CH: So, they are all making the Ascension?

MC: Yes, absolutely.

CH: Does a person just have to choose or request to make the Ascension? What is the process?

MC: It is a choice, it is an understanding, it is a knowing that they belong together and that they are being helped by the Light to do this all together.

CH: Is that why we came, to make this shift?

MC: Absolutely.

CH: What can we say now to the people of Earth to give them inspiration, hope, and courage to get them through these last days?

MC: It's all good, it's all in Divine Right Order. Be happy and be at peace. Peace in your heart, peace in your whole being. Peace, peace, peace.

CH: All that has been written shall come to pass. This is the law and it is unfolding in Divine order and Divine timing. As it was in the beginning it is now and ever shall be.

Chapter 9

Help Comes from the Stars to Make This Dimensional Shift

Susan Michelle Moll is a very powerful crystal healer and Lightworker, she is one of the ancient Lemurians who have returned for the end of the cycle when the Earth is being restored to wholeness, harmony, and love. Sue has a particular connection to the new children who are coming in now with a very high frequency. Her focus is to create new educational systems where these children can have the freedom to develop their natural gifts and abilities. Her Higher Self, Astara, agreed to answer some questions to help us to have more understanding of the coming Dimensional Shift.

Charmian—Let us go to December 21, 2012, and the alignment between the galactic center and the Earth. What does that look like?

Sue—I'm getting an image of the mother ships, lots of ships in a line, one above another holding the frequency for us. Then it goes down into the central core of the Earth connecting with the crystalline core. There is a sparkling crystalline grid; it's not around the Earth, it's inside the Earth. It's as if the ships are helping to energize this grid and I am seeing that this energy goes right through the Earth and out the other side; it just carries on.

CH: I wanted to get more information about the ships which S saw as I hadn't had that information before, so I did a session with my friend Colin to ask for more information and this is what we received:

CH: I am going to December 21, 2012. I am seeing a line of ships above the Earth all stacked one on top of the other. There is a beam of energy coming down from the galactic center, and the ships act like a series of filters and amplifiers of the energy. They take the energy in, and they add to it, refine it, and filter it. When it comes out it is more highly charged. The energy goes down from the ships into the Earth's crystalline core, and then there is a burst of light like a starburst as she sends it back up to all the people on the surface, all the Lightworkers. They are taking it up through their feet; it goes up to the heart and then into the crown chakra and pituitary gland. Do you feel that?

C: Yes, I got it then as you described it. There is a network going out from our hearts so all that energy comes up through us. It goes up through our crown chakra and back out again. We are radiating it out from our hearts to each other and then across the Earth; a network of beams interconnecting one another with this energy.

CH: You and I are receiving this energy now as if it were December 21; we have gone into it and are receiving it.

C: We just became One with the energy, it is just arriving.

CH: It's a preparation, it's familiarizing us with the energy so that when it happens we will be able to absorb it very quickly and then help other people to receive it without so much disruption because it is so powerful. We have a connection to the ships. I see that both of us have one of our aspects up in one of the ships. I am feeling myself here on the Earth but also in one of the ships standing at a control panel.

C: The sense I get is that, although the beams are coming out of the ship, it is the Beings in the ship who are focusing their heart energy into the beam. The energy is not coming from the ship's structure, it's coming through our hearts.

CH: There is a circle in each one of the ships. Each circle receives the beam and adds their quality to it, then passes it on to the next one. We are in one of the circles, not in the first ship but about five down. We are in the circle that is receiving the energy, which we pass through our bodies, refining it and adding our qualities to it.

C: It feels like there are others in the circle who also have aspects of themselves on the Earth at the same time. We may not know them on Earth, but we know them in the ship. Here I feel as if I have known them all my life.

CH: There is a very strong connection. We are standing in this circle and each of our hearts is a star that is pulsing; each pulse connects it to the others in the circle as if we are one heart beating together. We radiate the pulse out to each other. It is the Christ energy we are pulsing in this ship; it is a Sirian ship. We know the others as a pod even though we don't know them on the Earth yet. This group is adding the frequency of the Christ to the beam. This is a very highly evolved group of Beings. I think that when we meet them on the Earth they are going to be very evolved spiritually.

C: There is a soft quality to the energy; it is very pure and very fine.

CH: All of the Beings in this ship may not be human; some of them could be cats, dogs, or dolphins on the Earth plane. I can feel the energy flowing through my physical body now, soft white and very gentle.

C: It's a very tender feeling like being cherished and cared for.

CH: Now we are sliding down the beam through the rest of the ships. One has pink energy from Venus, the next is gold, and then the last one is aqua blue. Each different ship is putting a different color into the beam of light. You and I are flying down the beam into the Earth's core in the form of our Sirian Christ bodies. We just hold the energy—the balance of male and female energy there in the center of the Earth's crystalline core.

C: She welcomes us as we join her in her center.

CH: This is the energy that Yeshua and Mary Magdalen brought to the Earth, the balance of the Female Christ with the Male Christ. When their lives here were complete, they left but now the energy is returning.

C: I felt them come in very strongly as you spoke their names. I just heard, "Know who you are."

CH: The energy is very soft, but it is huge and very powerful.

C: I felt every aspect of myself out as far as you could go in all time and space. I felt them all come together as One. All of the different aspects on the planets I have been on, all the forms I have taken on all of the different planets I have lived on. That is the recognition of Who

We Are. It sits well within the energy. This is the energy that is creating the New World that we want to live in. I am seeing it open up before us, seeing the beautiful green planet and how she responds to receiving this Love. She is glowing. It is a complimentary cycle of creating and giving back, a natural cycle. We give to her and she returns our Love. We are all co-creating this together. We need to gently bring it into our DNA structure.

The rest of the session returns to S and the session with her Higher Self.

CH: Let us ask for the consciousness of the Beings on the ships to join with us to help to prepare us to make this Dimensional Shift. What happened when this energy was fed into the global crystalline grids on the surface of the Earth?

S: It was an activation for the people who are living on the surface of the Earth—an awakening and a remembering in different degrees of their awareness. Each person was affected by it. There was the power of this activation and connection on that day. It had already started, but it was increased to the highest vibration that humanity can absorb at that time. It's as if everybody walking on the surface of Earth has a point of Light connecting from the Earth to his or her feet. It's coming from the Mother. It's like we are being beamed from Heaven and from the Earth to connect to our hearts. The connection is then made from the heart to the mind. It comes down through the Earth then back up through humanity's feet and chakras, activating them at a cellular level.

CH: What effect will this have on how we live on the Earth?

S: It will automatically have an effect because everybody will awaken to a certain degree and the effect will increase, but it will be a gradual process. People will automatically begin to shift and change and look for higher and better solutions. They will be brought into contact with these solutions as they open up to receive them.

CH: Will this energy cause any geological events such as earthquakes or tsunamis?

S: I'm not feeling that this is going to happen. It is a very positive experience—it's a gift that is giving everybody on the Earth a chance to awaken even if they have been in a deep slumber.

CH: How will it change the way we are living now? How will it be different?

S: It will be a gradual process. It won't just happen overnight. People will begin to live more and more consciously day by day as their frequencies rise. They will automatically awaken to the higher frequencies. Their hearts will open more and more and there will be an automatic choice; they will choose the highest good for themselves and for their families.

CH: What about technology? What will become available to us that we are not using now?

S: This technology, as we all know is there, is available. It's a matter of not imposing this technology on humanity but of them opening up to receive it. This process is an empowering process. It's not something that the galactics are going to drop into your laps; it is something that they are guiding us all to do. This is a process that humanity, with their help, is bringing forth—the transformation of themselves, the transformation of the Earth and all of the technologies. The knowledge is all here ready and waiting.

CH: Do we need to ask for it?

S: There will be certain individuals who will bring forth this information and you will have everything in place. They will be the ones that have been guided all these years, learning how to bring this new technology forth upon the Earth. They will do this, and they will bring together the groups of people that will assist them in this process.

CH: What about the technologies we are using now that are causing so much pollution like gasoline?

S: They will all disappear. It will be a gradual process, but they will all soon disappear.

CH: How will people organize themselves politically in terms of power structures and government? How will they govern themselves?

S: There will be whole new ways of society governing itself. There will be groups within groups within groups, circles within circles within circles, and everybody will have their part to play. One particular person, or just a few people, will not rule over an entire country. Governing will be done by pods of people who decide over their own community, their own area.

CH: What about this plan that some people are talking about to impose a world government and to enslave this human race. What will happen to that plan?

S: Everything like that will be dissolved.

CH: So what can you say to people who are in fear or who are worrying about this supposed plan? What can you say to them to help them to move to a better perspective?

S: Raising their consciousness to a higher frequency is always and will always be the solution.

CH: How will we be relating to other kingdoms on the Earth, such as the animal kingdom and the plant kingdom, after this awakening?

S: This will be a marvelous time. Everybody's senses will begin to sharpen; their sixth senses will begin to awaken. They will be more conscious of the elemental energies, of the fairies and their interaction with the plant kingdoms, the mineral kingdoms, and the beings of the sea. Humanity will gradually begin to see the wonders that they have, for so long, not been able to access because of the frequency difference. When we have all raised our frequency, it will be as if the veil is drawn aside once again and we will be able to connect with all the kingdoms that we have been missing for so long. It will be a wonderful time.

CH: How will the children be in the new world?

S: The children will be happy, finally, and not just the young children but the older children, the grown-ups, the ones who have been waiting for so long. They will finally see why they came and will be able to implement their purpose after waiting for so long.

CH: People like you and me?

S: Yes, and our children; I was seeing our children just now. We are like wisdom keepers. It is not so much that we are going to be actively involved in the transformation; we are going to be more guiding the younger ones. They will be the ones who will be very active, and we will be guiding the process.

CH: More like holding energy?

S: Giving them the guidance that we receive.

CH: Why are they having such a hard time now, these young ones? They are being challenged so much—why is that?

S: They too have their own imprints to dissolve and to integrate just like everyone else and because they were born when they were they still have those imprints.

CH: Is there anything we can do to help them to clear the imprints?

S: Clearing our own as much as possible and assisting them by sharing everything we know with them if and when they are open, but not to bombard them with too much information. When they are willing, just share our clearing and integration with them. They are automatically involved with it and it will help them. It ripples out like a stone in water. The young ones, our children, some of them have become very disillusioned because it is taking so long, a lot longer than was originally planned, and we have felt that too. They needed to have their experience. Everything they have experienced was part of their imprinting with what they asked to learn. We must trust that they know and are guided just as much as we are. Don't worry about them but hold them in the highest Light and always see them flourishing, radiant, and learning their lessons just like we have done. There is nothing to fix, nothing to change, it's all perfect. They are clearing and integrating just as we are, so whatever the lessons are, it's all perfect. There is no need to take on their clearing and integrating as well. When they ask for guidance, we can give it to them and assist them in any way we can.

CH: Is there anything that people in general can do to help with this shift?

S: Connecting to the Earth is vitally important—connecting to the crystalline core. Stay grounded to connect Heaven to Earth and bring the energy back up so that they are already tapping into this band, this crystalline grid. It is already activated so if they become accustomed to the frequencies by being in nature, being in the ocean, cleansing the body, drinking lots of water, and remembering that they are Love, this will help. By remembering that they are Love. Love everything, all the feelings and all the emotions; feel them, integrate them, and then the triggering will dissolve and be erased. The more they can stay in the band of Love in the higher frequencies, the easier life becomes. It's not just about ignoring anger, pain, or suffering; it's about feeling it and integrating it, allowing it to dissolve. It's not about rejecting the dark side—which is in effect not the dark side, it is also God—embracing and receiving everything as part of the whole. That is what we need to do. Embrace everything, for it is all part of the whole.

CH: How will we relate to the Star People and their ships after the shift when we are in higher consciousness?

S: We will be able to communicate with them and, eventually, be able to walk and talk with them. This will also be a gradual process because if it happened too suddenly people would be afraid according to their awareness. As it is now, some can see and some can't, some feel and some get messages. The connection will be much stronger, particularly for those who are guiding the process.

CH: Let's look at the ones who are most ready to make the shift and who are raising their frequency. Let's look forward to what the transformation will look like for them.

S: It's more people than we think. I keep seeing and feeling how the energy just keeps connecting, going round and round in the vortexes, as it is imprinting us in every single cell, gradually day by day. It's happening now. It keeps going from one to the other; it is passed on.

CH: So the people who awakened first will activate others?

S: Yes, it's like they pass the baton to the others. It is an energetic thing. I'm seeing streams of Light connecting to people. They integrate it and they pass it on. It's nothing that we have to do, it's automatic.

CH: How do the children fit in? Are they more receptive to these energies?

S: They are at the core of this.

CH: Is this all the children?

S: All the children; however, some are a little denser, but mainly I see the children are the core of this; they are doing it already. It's like a hub of energy where the children are the key point.

CH: Especially these new children that are coming in now. So how can we support these high vibrational children? Is it difficult for them right now living in this frequency of density?

S: Sometimes. It is the same for them. If they manage to stay in the higher frequencies they are fine. If they go into a lower frequency and dense places, they find it very disturbing.

CH: Is this something parents need to be aware of?

S: Definitely yes. Places like Walmart and anywhere there are lots of people and lots of electromagnetic frequencies. They have a really hard time, it pulls them off balance and it does us too. So avoiding these places and being in nature as much as possible, connecting with the ocean and being raised by ones such as us will be most beneficial. The support of people who are awakening will help and they know who you are. This is very important for them.

CH: If they or we are exposed to high electromagnetic frequency (EMF) waves how can we restore balance?

S: Having a shower or bath, going into a pool or ocean can immediately cleanse the effects. Calling upon the crystalline flame to dissolve all distortions in the energy field will also be helpful.

CH: After we move through December 2012 and the Light is rising, what happens to the old structures that are in density and to the people who choose not to make the shift?

S: I'm seeing that it all gradually dissolves.

CH: Is there anything that Gaia needs from us now to help with this process?

S: She needs our connection; connect daily with her core, bring in the Light, and keep the connection going.

CH: What role do the whales play in this transformation?

S: They are holding the grids around the planet. They are helping to hold the planet in balance as much as possible because they are in such a high frequency. They don't get into lower vibrations like we do, they manage to stay in the high frequency. They don't think like we do.

CH: What is the role of the dolphins?

S: They are helping with this too. They are bringing back the element of joy, ecstasy of living, vibrancy, and aliveness that we have been missing through the time of density and oppression that we have been living through and all that we have experienced. They are holding this higher frequency for us so that we have something to anchor into. They are telling me to anchor into their frequency. If you are feeling really down, just ask them to elevate your frequency to a higher vibration and they will joyfully assist.

CH: Can you do this by thought? You don't have to be in the water with them?

S: Just ask them to assist you; they play a big role in this.

CH: Is there a message for Sue from Astara?

S: Continue to embrace herself and to love herself where she is right now; to trust that everything is unfolding perfectly in Divine timing. She (Astara) knows that it has been challenging, and she says I have been doing well, better and better. She says just to keep my heart open and to keep on keeping on and to keep my frequency high. Not to allow myself to sink down into trying to fit into this world that doesn't fit. Just to trust that I am being taken care of.

Chapter 10

Emergence, Inner Earth, and Beings from Telos Come to Assist

*B*eloved Love is a singer and composer of celestial music. Her music carries codes for activating the heart and the Divine Feminine energy. In her session she was able to connect with the Lemurian Light city of Telos, a community of Lemurians who went underground at the end of Lemuria, that is located underneath Mount Shasta. They are assisting us on the surface of the planet to prepare for the Dimensional Shift, after which they will return to the surface of the Earth and bring us new technology to help us to build the New Earth. She also connected with the civilizations of the Inner Earth that are very highly evolved and are also waiting for us to move into connection with our God consciousness. When we have stopped the wars on the planet, they will come and help us to clean up the toxicity we have created on the Earth.

Beloved is spreading the Light throughout the planet through her music. After the Dimensional Shift she saw the New Earth with all people living in peace and harmony.

Charmian—The Earth is coming to a time of great transformation. Let us go forward in time to see what that looks like as we come closer to the time of the Ascension. See yourself moving forward through time to where there is a great change upon the Earth. Tell me what that looks like.

Beloved—I see positive people sharing with great love and respect.

CH: How is it different to the way it is now?

Be: There is no mental illness, no disease, and there isn't even any money.

CH: How are people supporting their physical bodies? How do they keep themselves alive?

Be: They are growing their food.

CH: How is it different from the way we grow our food now?

Be: People are growing what they want, not to sell, just what they are drawn towards for their own nourishment that everyone can share.

CH: Let's look at the consciousness of the people on Earth. How is it different from the way it is now?

Be: There is no fear.

CH: What are their abilities in terms of consciousness? How do they communicate with each other?

Be: They can talk and they are telepathic, but they like to sing and dance too.

CH: Let's look at the telepathy. How do they communicate with the other kingdoms, the animals and plants?

Be: They can speak either out loud or communicate with telepathy.

CH: How do they connect to the spirit of the plant when they are growing their plants? Do they do something with the essence of the plant to help it to grow?

Be: They show love, appreciation, and gratitude.

CH: Can they communicate with the animals?

Be: Yes, they communicate telepathically.

CH: Tell me what happened to the Earth to create this transformation, was it an event that triggered the shift?

Be: It was time and enough people wanted it.

CH: If we looked at the calendar what would be the date?

Be: Saint Germaine says July 25, 2013.

CH: What did people do to prepare that for this shift?

Be: They worked daily with the Violet Flame. Singing, chanting, yoga, and healing arts all helped. They came together in festivals and gatherings that brought great joy and upliftment. There was a feeling of invoking the Oneness of All Beings and All Things.

CH: How many people made this transformation?

Be: They are still deciding; some people are still making the decision.

CH: What happens to the people who choose not to make the shift?

Be: They create their own reality that is for their highest good and spiritual evolution.

CH: What does that reality look like?

Be: If they still want war they can do that.

CH: How do those who do make the shift and those who don't make it experience the New Earth after the shift?

Be: The people of the New Earth who have made the shift are sending Light and assistance to the ones who didn't make the shift.

CH: How are they assisting them?

Be: The same way the Ascended Masters are assisting us now. Those who long for the Light shall receive it according to their ability to receive.

CH: Imagine we are looking down on the people of the Earth after the shift. How are they living?

Be: In community. They are sharing their food; they are eating live food and drinking pure living water from the springs. Live food and living water are cleansing them of disease and aging.

CH: What kind of structures are they living in?

Be: A material you can see out of but people can't see in, so it maintains privacy while giving a view.

CH: What is it made from?

Be: It is made from plasma in a molded form.

CH: How did they know how to make that? Where did the technology come from?

Be: It came from inside the Earth, from Middle Earth.

CH: What happened between humanity and other civilizations at the time of the shift? Did the people from Inner Earth come to help us?

Be: The portals were opened because the humans on the surface of the Earth were having positive thoughts all the time.

CH: What do the Inner Earth Beings look like?

Be: In Inner Earth they still look human, but deeper in towards the Earth's core the Beings are fifteen feet tall, more like Avatars.

CH: What would be their message now for those of us who are preparing for Ascension?

Be: Be in pure Love and joy. Gather together in singing and dancing, sharing food and celebrating life.

CH: Are Beings from other star systems helping with the Dimensional Shift?

Be: They are bringing technology to clean up the planet and each sharing their unique expression of Love.

CH: Did something happen to humans that made them more receptive to hearing and seeing Beings from other dimensions?

Be: First, they knew that it was possible, and then through their desire and willingness they were guided into contact with other civilizations.

CH: Look at the bodies of the humans after the shift. How are they different?

Be: They are more light than dense, more in their Light Bodies.

CH: Where does the Light come from?

Be: From the heart; from their own God consciousness.

CH: How is life different from the way it was before the shift?

Be: It's not about survival anymore.

CH: How do people get the things that they need to live?

Be: They manifest what they need by pure intention.

CH: Is there a specific practice or ceremony they use to be able to do that? Is it a group or an individual process?

Be: Some are taught individually and some in groups.

CH: Ask to be shown the process of bringing things they need into manifestation.

Be: By visualization and unwavering concentration it is brought forth in Divine timing.

CH: How is this connected with the memories of Lemuria?

Be: That is how it was done then. That was the seedtime for the coming Seventh Age of Gold.

CH: What is the significance of Hawaii in this process?

Be: Hawaii holds the Codes for returning to Oneness. Ancient Lemuria was a beacon of Love in this galaxy.

CH: How are the Codes to be spread from Hawaii to the rest of the planet?

Be: Simply by developing them; they are being seeded in the planet.

CH: When Beloved takes her music out will she be assisting the Light Codes to open in others?

Be: Definitely, that is why she is here on Maui now, to receive the Codes.

CH: Would being with the whales and dolphins help?

Be: Yes, being in the water with them and hearing their song.

CH: Let us see the Codes opening now. Ask to be shown the DNA spiral and the Lemurian Light Codes within the spiral. What does that look like?

Be: The helix is spiraling up the Ascension channel in both legs and then into the hands and the heart. It is like a doughnut shape or the Fibonacci* series shape.

CH: Let us see and know that the Light Codes are being activated from this time forwards. The frequency will be in everything that Be does from now on—sound, dance, yoga, everything. She will be carrying the vibration wherever she goes and it will be going into her CDs and her music. She will be transmitting the frequency as a priestess of Mu. Since the time of Mu many more layers have been added, so the energy is more multidimensional. Look at the complexity of the geometry of all of your aspects coming together as One. What does that look like?

Be: Whales breaching.

CH: Feel that in your body now, everyone you have been on all planes, all dimensions, through all time, through eternity, past, present, and future, are all assembling here right now. They are all being downloaded into your energy field, all working together as One. We see this in the name of the One and we give thanks that it is so. I am seeing a spinning, multi-layered geometrical matrix in your heart, spinning all of its colors. It has jewels on it and it's like a control panel. It has all of the activation programs on it for Ascension and it is slowly turning. It is like a communication device from the Divine director between all of the multidimensional layers and aspects of your Being. In this New Earth you have connection with all of the planes of reality and with your twin flame, who is now much more accessible.

CH: Let us ask what your connection with your twin flame will be when you go through the Ascension?

Be: We will decide when we meet.

CH: Will you be on the same level of physicality?

Be: We will both be in the fifth dimension.

CH: What happens to the children at the time of the Ascension?

Be: Those who are ready will ascend.

CH: Is there anything we can do to help those who have chosen Ascension to prepare?

Be: Guided meditations, reminding them who they are, honoring them as Ascended Beings.

CH: I have just received guidance to make healing journey CDs for children who are sick. Why are so many children manifesting serious diseases like cancer at this time?

Be: It is to help their parents become more aware of the toxicity of the environment, habits, patterns, and beliefs.

CH: How are communities in the New Earth organized? Are there governments or are communities self-regulated?

Be: There is no need for governments. Everyone knows their own natural tendencies and abilities.

CH: So how is harmony maintained?

Be: Everyone is in harmony with their own Divine Presence and that brings harmony to All.

CH: Do you see a connection between the New Earth and the Lemurian city of Telos underneath Mount Shasta?

Be: It is heavenly.

CH: Do you see Adama and Lady Galatia, the high priest and priestess of Telos? What message do they have for us at this time?

Be: Open your heart and be at peace.

CH: What will happen with Telos at the time of Ascension?

Be: We will be reunited.

CH: When does this happen? Are there stages of Ascension and what is the significance of 2012?

Be: It is a galactic alignment and an astrological alignment through the planets.

CH: When we aligned with the galactic center in December 2012, what happened on the Earth?

Be: I see the Light bursting from the core of the Earth and radiating out.

CH: What effect does that Light have on the surface of the Earth?

Be: There is peace and everyone knows their God Self, their "I am" Presence.

CH: What happens to the people who choose not to ascend?

Be: They are in another place, another planet.

CH: How does that work?

Be: I see two Earths, but that doesn't have to be so, there is still a possibility that everyone can ascend.

CH: If they make that choice?

Be: Yes.

CH: Are there events scheduled to happen on the Earth to help people to make that choice?

Be: Every moment the choices that they make will decide whether they further their spiritual evolution or not.

CH: Is it important for people in karmic relationships to complete them now?

Be: Yes, it is important in order to most efficiently bring the Ascension frequencies in as quickly as possible.

CH: Is there anything else that Athena, your Higher Self, would like to say at this time?

Be: The particle electrons in Beloved's body are truly in balance at the moment.

CH: So we bless you now with the grace and love of Divine Mother. Be the beauty that You Are, be the Grace that You Are, and be the Love that You Are.

Fibonacci series: In mathematics, the Fibonacci numbers or Fibonacci series or Fibonacci sequence are the numbers in the following integer sequence:

$$0, 1, 1, 2, 3, 5, 8, 13, 21, 34, 55, 89, 144,...$$

By definition, the first two numbers in the Fibonacci sequence are 0 and 1, and each subsequent number is the sum of the previous two.

Chapter 11

The New Way of Living from Joy

Keala Gerhard is a very gifted healer. She works with Reconnective Healing, a technique that helps people to connect to their higher frequencies. This work helps them to move through the energetic changes on the Earth more smoothly. I guided her in hypnosis to a time in the future when the Earth had already gone through the Dimensional Shift. The time she chose to move into was quite a long way forward to 2035 when the changes were well established and everything on the Earth was peaceful and settled.

Charmian—Let's go to the time when you have gone through the Dimensional Shift.

Keala—I feel myself outside in nature. I am walking between buildings. There is one central building and there are other buildings around it where we live. The central building seems like a temple or a meditation hall.

CH: What does your body feel like?

K: It feels very healthy. I can't feel my age and I feel strong and young.

CH: Who else is with you?

K: There are men and women; we live in the houses around. Each person has his or her own room. We have partners but in a different way, more like friends. It's not something that feels limiting, it's very open and light. It feels very loving and connected as if it is easier to find your soul mate. You trust in guidance, you care for each other, and you take care of each other.

CH: Do you have the support of the community?

K: Yes.

CH: What are you doing in the community? What is your function?

K: Very light work with crystals and sound.

CH: Do you and your partner work together?

K: We are partners even though we do totally different things. He does art or architecture; I do healing.

CH: What is the energy like between you?

K: It is absolutely supportive and light; there is so much freedom. We share certain activities like reading poetry to each other or going out together. I don't even cook!

CH: Do you eat?

K: It seems like we only eat fruits and nuts; only light food like liquids. There are no kitchens.

CH: Why don't you need to eat very much?

K: We get our energy from the air, from the light.

CH: What is happening on the rest of the Earth?

K: It is very peaceful and not overpopulated. It feels as if there is not much noise or technology.

CH: It sounds like life is much simpler?

K: Yes.

CH: What year would this be?

K: 2035.

CH: How did the shift happen? What was it that shifted us out of the old heavy dense energy?

K: It's about vibration and frequency. Our bodies were being prepared; our nervous system was being worked on. It was shaken up like a rattling. Shaking off all of the old past stuff. This is happening in our time on the Earth now, it often feels too much.

CH: How can we help this process so it doesn't feel so overwhelming?

K: Walking on the Earth is good and swimming in the ocean. All of my insides are shaking and there is a feeling of urgency and time; we don't have much time.

CH: Let's look at the sequence—all this shaking, the Earth frequencies are rising higher. What is happening underneath as the old stuff is shaken off? How is the body underneath the shaking?

K: There are new channels, new DNA, new crystalline DNA strands are being added to the two old ones.

CH: The third strand, the crystalline one?

K: They come in threes: 3, 6, 9, then 12. The third one is crystalline; it is encoded with the Oneness frequency.

CH: What stage is Keala at now?

K: The third one is building.

CH: Let's go to the time when the Light Body is fully activated. Tell me what that looks like?

K: I can see and feel that the nervous system is really stable, it feels clear and solid, but at the same time luminous. It is clear like a crystal, solid but luminous.

CH: How long did this process take until the full activation of the Light Body?

K: It was in 2012 between the middle of the year and the end of the year.

CH: Is this something that many people experienced at the same time?

K: Yes, about half of the population received the activation.

CH: Is it a certain frequency that the Earth reaches that triggers the Light Body activation?

K: Yes.

CH: As people are being shaken so that all the old stuff comes up and releases, how does that translate on the global level in terms of countries and nations?

K: It depends how things changed in the first part of 2012.

CH: Do you mean how the consciousness of the people changed?

K: Yes, whether they go with the new frequency or they choose not to, it's mostly about forgiveness. If they stay with their old patterns of hate and revenge, if they don't want to change, they will destroy themselves. If they choose the path of forgiveness and Love, then all will be well.

CH: Is it important for people who are evolving to be in certain places while the process is going on?

K: Yes, they will be moved or they will feel an inner call.

CH: Let's look at the Earth after the Dimensional Shift. What does that look like?

K: There are fewer people, more greenery, and more clean water. I see pictures like out of the movies—like springtime—everything is so beautiful. There are scenes of people growing more simple kinds of things. There is no sophisticated machinery; everything is more natural. The houses are simple. There is machinery but it is run on solar power; there is no noise. Everything looks more beautiful. People are laughing and happy. They are working together and supporting each other.

CH: What is the connection between animals and humans?

K: The connection is very loving. There are more animals that are beautiful such as cats, dogs, birds, and horses, rather than the animals that we eat now.

CH: How do we communicate?

K: They understand; we can talk to them. They understand the vibration, and they can tell that there is more love just from our expression.

CH: What are people doing with their time?

K: I see them coming together in the late afternoon and evening. I see them dancing and being creative; they read poetry, sing, and dance together. There is no TV.

CH: How is society organized and structured? Is there a government?

K: There is a beautiful big round building, six sided, and it is in the middle of the community. Many people come together there; they meditate and then they discuss things and come to solutions.

CH: So when there is an issue or a problem, how is it resolved?

K: They all meditate together and receive answers, then the highest thought is chosen which is for the good of all.

CH: Is there any money or currency?

K: No, I don't see a currency. Everybody is provided for; there are big stores and everybody does what they like to do. There is something like little pearls that people have; they carry them around their necks as exchange.

CH: What kind of activities do people do?

K: Everything creative; there are woodworkers, weavers, fishermen, and healers. Everybody does what he or she loves to do, and they also do communal work. There is much more free time, they only work half a day.

CH: How are the children raised and taught?

K: There are big schools right in the middle of the community. The houses are around the center. The children are taught more through play, like Montessori. There is a lot of creativity and play, lots of nature; they learn a lot by doing and experiencing and by being outside. I can see glass rooms; almost all glass and their windows can be opened. The children go out in nature a lot.

CH: How is healing done?

K: It's a lot of healing with light and sound and hypnosis. There are healing chambers, special healing temples or centers. They use a lot of Light.

CH: What is Keala doing with her time in this new world?

K: I have such a long history with hands and healing. Some people like to be touched and some don't.

CH: How is your body after the Light Body is activated compared to how your body is now?

K: We move so much and our bodies are strong. We dance and walk more; we are so alive. We express our feelings, we laugh, and we cry. We are happy, we are healthy, and our inner system is stronger. We have counselors and if we have problems we are free to talk about anything. We are all so ready to express our feelings, so there is not much illness. I don't see any sick people.

CH: What age does this new body feel like?

K: It feels more like thirty or thirty-five.

CH: What can you tell Keala to help her between now and the Ascension time?

K: She is on the right track. Trust more, rest more, and rest for ten minutes every hour. Trust, just trust, relax, and have fun. She has been taking things very seriously and it isn't helping.

CH: Do you have a message to all of the Lightworkers?

K: It's really about taking things more lightly, being joyful and celebrating. Be happy even in your sadness, fear, anger, and disappointment. Love it. Hug yourself. The destination is already set, so whether you go to it in Light or whether you go dragging a heavy chain, is up to you so be gentle with yourself.

Chapter 12
Releasing Old Karmic Relationships

A came to a crystal bowl and Mother Mary channeling event I was doing on the islands and seemed very subdued and unhappy. She was exhausted from many years in a challenging and unfulfilling relationship with a man who was unhappy with himself.

The message from Mother Mary at that event was all about self-nurturing and self-love. She said that many of the people in the group were "givers," and they needed to focus on also giving to themselves. She encouraged them to ask for what they need from the people there were giving to. The people around us don't always notice when we are in need of support and nurturing. We have to ask for the help we need and also to give it to ourselves by engaging in activities that feed us emotionally, physically, and spiritually. "A" had been in a karmic relationship for many years with a man who was emotionally disconnected and was unable to give her the nurturing that she needed.

Many people are in a similar situation at this time on the Earth. This is the last opportunity to fulfill karmic debts with those whom we have unresolved karma with from our past lives. We come together with those by whom we have been disempowered in previous lives. We do this in order for us to reclaim our power from them or to give them the opportunity to repay whatever karmic debt they owe to us. Karma is ending now on this planet, so those who still need to complete karma will have to do it on another planet or ask to receive special dispensation by Divine Grace to be released from it.

It is time now to be released from these karmic bonds and relationships as the time comes closer to the Dimensional Shift. We need to be connecting now with those we have contracts with to make the Ascension together. There is no blame or failure; many of us have been through several of these relationships in this lifetime. It isn't a weakness or anything we are doing wrong, it is just that this is the last lifetime for karmic completion. If we have more than one person to resolve karmic issues with, then we have chosen to take this opportunity to find resolution and completion. These relationships are very challenging and the time for them is fast running out. Another way to resolve the debt, rather than having to go through each relationship with unsure break up and conflict, is to unconditionally forgive and release any and all beings who have a karmic debt to you or to whom you owe a karmic debt through all time and space. You can make a ceremony for yourself to achieve this. You don't have to be specific about who they are, but you can also name and forgive each one for those you know about.

Charmian—Let us go to the time after the energy shift on the Earth. Where are you?

A: I am in nature.

CH: How does that feel?

A: It feels safe and healing,

CH: What is the quality of the light on the New Earth.

A: It is bright and beautiful. There is such a vastness, as far as the eye can see. It goes on forever.

CH: What does the body feel like?

A: Very light and airy.

CH: How are your emotions, your feelings?

A: I feel very present; I am feeling Love. There is a feeling of gratitude, and there is a feeling of sadness.

CH: What is that connected with?

A: It is sadness for what was. I feel it in my solar plexus and in my heart.

CH: Is it to do with other people?

A: It has to do with others but also the child within myself.

CH: How is she feeling?

A: Alone.

CH: Divine Mother is in this Light with you in this New Earth. There are no limits, no walls, and no barriers. All beings are completely accessible to you, so call your soul family, your kindred spirits, and your spirit helpers to you.

A: First, I see the child running, smiling, arms wide open. Mother is waiting with her arms wide open.

CH: Which form of mother is this?

A: My Guardian Angel.

CH: Feel yourself being enfolded in perfect Love. The sadness, the pain, is being left behind.

A: It is gone now.

CH: What does the New Earth look like?

A: It glows like the Golden Sun.

CH: How are the people on the New Earth different to people now?

A: They are all open-hearted and it's as if they are floating, they aren't even touching the ground. I hear their laughter, and I feel the warmth of their love.

CH: How are they living? What are they doing?

A: Very simply, there is a great sharing, there is no lack or need, there is no waiting for anything. It's all there.

CH: Look for your own family, your soul group. What does that look like?

A: I see my grandmothers. We are so happy to see each other. My heart feels so overjoyed. There are women and children coming from everywhere.

CH: What is your purpose with the women and the children?

A: To fill them with Love and Light; to heal them.

CH: How do you heal them?

A: With joy and love and acceptance, I teach them to embrace All That Is and to fill it with love. There is such gratitude.

CH: How do you sustain the body in this New Earth?

A: With continuous love, feeding it what it needs—truth, light, water, and breath.

CH: Is it still a physical body?

A: It doesn't feel like the physical body. It feels very light.

CH: Are you living in any kind of structure?

A: I am out in nature. There are places where I go to stay. It is in the nature of All That Is.

CH: How do you feel towards all the other kingdoms of nature? How is your consciousness connected to them?

A: There is a kinship, no division and no separation, just wholeness.

CH: How are people relating to the animals and birds?

A: We are relating to them as we are relating to each other, there is just a harmony.

CH: Can you tell me the connection between Lemuria and this New Earth?

A: In Lemuria there was a deeper definition; it is the same in the New Earth. The heaviness is gone, the darkness is gone, and the murkiness of it is gone.

CH: What happened to the darkness and the heaviness? Where did that world go—the world of suffering, pain, greed, and war?

A: It just disappeared. It went deep, deep into Mother Earth. It was a cleansing and it was purified.

CH: Did something happen to cause the Old Earth to disappear? Was it a particular event?

A: It was a releasing, a surrendering.

CH: How do the Hawaiian Islands relate to the New Earth? Why has A been called to the Hawaiian Islands?

A: To show her how it was. Watching the water as it ebbs and flows so effortlessly, the wind dancing with the palms, the leaves, the trees, the flowers; for her to see the gracefulness, the openness and expansiveness of it all.

CH: Is this connected to the memories of Lemuria, being on the land and in the water in Hawaii? Is this helping her to remember?

A: Yes, I am feeling the whales; they are singing in the water.

CH: How does the whale song help A to shift into the new frequencies?

A: I feel the water as an energy vibrating inside me, starting at my feet and working its way up. It is very electrifying.

CH: What message does your Ascended Self have for you?

A: Nourish and love yourself. Be free.

CH: How can A hold this energy through the time of waiting between now and the energy shift on the earth?

A: Be as you are now.

CH: How can A bring this energy into her current relationship? Look at your husband's energy field. What does it look like?

A: Very dark and heavy.

CH: Why is that?

A: He is very alone, he feels very alone. He is lost, very lost. He hungers for the very thing he already has. He doesn't see it.

CH: Is there anything we can do to help him to remove the veil from his eyes?

A: Surround him with Love, with a beam of Light to shatter away the hardness in his heart.

CH: Is it really himself that he doesn't like?

A: Yes.

CH: How can A maintain her Light energy in her husband's field of darkness?

A: She must give herself permission to shine brightly under any circumstances. She can shine brightly and let Love and Light filled up all the crevices.

CH: Let us call in her husband's Higher Self. What does that look like?

A: I see the warrior of the woods. He is trying to connect with the personality self who is so wounded.

CH: What is blocking that process?

A: He is very wounded.

CH: Let us as ask Divine Mother Mary to take him, to hold him, and to care for him. Let us ask for you to be released from being responsible for him. Mother Mary says, "As you see in the Pieta (the image of Mother Mary holding Yeshua's body after the crucifixion), I handed over the body of my son to father God, so you must hand over the wounded body of your husband to me. I will take him and I will hold him and you are free."

CH: Is there one last message from A's Higher Self at this time?

A: You are whole and complete. All is well.

Chapter 13
Awakening to a New World

"An" is another old soul who was visiting Hawaii. In the session she had very clear recollection of her soul's purpose in coming from the Light to lift the Earth and humanity up from darkness into Light. Everything that is happening on the Earth right now is part of the Divine plan to fulfill that purpose and that destiny. There are many Lightworkers who are awakening to their own power and the truth of who they are as co-creators of the New Earth. I guided "An" through the time of the Dimensional Shift and asked her what she saw.

Charmian—There is a doorway. On this side is the Old Earth where there is suffering and pain. Through the doorway is the New Earth, the Healed Earth. Go through the doorway and tell me what you see.

An—I see the sun, the water, and the vegetation. There is just a harmonious feeling and some people just come together, there are no traumas, no ups and downs.

CH: How does the body feel?

An: It's exciting. "What shall I do now because I can do anything I want?" It feels like everyone is family.

CH: What can you do in this body that you couldn't do in the old one?

An: I can be either here or somewhere else just on my whim.

CH: How do you move?

An: I like to walk but I feel that anything is possible.

CH: How are people treating each other?

An: They are acting like "what a pleasant surprise!" It is like everything and everyone just realized that something has changed, like they just woke up.

CH: What kind of structures are people living in?

An: I don't see structures. Everyone is moving around and looking, like it's so exciting and they are starting to check out where they are. The structures are very harmonious and there are no corners. They seem to flow off the landscape and are structures made of all curves—no abrupt shapes.

CH: What are people doing now? What is their focus or their purpose?

An: They are just checking in with each other.

CH: What does the Earth look like?

An: It's softer; there is more growth, more trees, more vegetation. It feels spacious, not cramped.

CH: How do people communicate with each other?

An: It's as if they know right away exactly how everyone feels.

CH: Is there more telepathy, mind to mind?

An: Yes, their voices are more just for singing.

CH: So what is the feeling on the planet as a whole and in the human race?

An: They are like the new children coming in who already know what they are going to do. Nobody has to tell them, and there is a whole lot of anticipation and excitement.

CH: How do you sustain this body?

An: I think without a clock, there is no time now.

CH: Do you still eat food?

An: Yes, it still drips down my chin and my hands. My hands still get dirty if I put them in the dirt.

CH: What about the animals? How do humans relate to the animal kingdom now?

An: It is like they really love the environment and everything that is with them in the environment. They are happy together.

CH: Do people communicate with the animals?

An: They appreciate the animals for how wise they are.

CH: What is the relationship now between the human and the animal kingdoms?

An: It's a lot more loving. Animals are more able to communicate with us, and we appreciate more how much they have honored us. They always wanted us to know that, but they didn't have a voice.

CH: How were the animals helping us before the Dimensional Shift?

An: They were able to stay grounded because they didn't veer away from their essence. They have helped us to keep grounded so that we could come back to knowing who we are. They have unconditional love, and they felt we were part of the animal kingdom too, a part that they didn't want to miss.

CH: Do whales and dolphins have a particular purpose in the transition from the old to the new way of being on the planet?

An: I think they have been continually holding a vibration here that has helped us to stay here. They have been dutifully doing that.

CH: So how does their energy shift when the Earth moves into Oneness? What happens to the whales?

An: Their song blends with ours now.

CH: What about the dolphins? What is their gift to us in this time of transition?

An: They are similar because they hold a certain vibration but they needed to have more connection with us. The whales were more like monks, but the dolphins were trying to get us to play.

CH: Was there something about the Old Earth that needed to have a change in consciousness? Did something happen that changed the consciousness of the human race?

An: Something very ancient but very familiar came back. It's like it was always going to be back and it's just the right time now.

CH: Is this to do with Lemuria?

An: I think it was before that. When we are in it, it will feel just like when we first experienced being on Earth, it will feel like we are back where we started.

CH: How are the tsunamis, earthquakes, and climate changes helping to shift the Earth? What is the purpose of these dramatic geological events?

An: It's a reflection of our own psyche, it just calms down as we get centered in our Higher Self. It all drops away like the parts that we are letting go of.

CH: So what happened to war, anger, and hatred on the Earth?

An: It's almost like it was just a dream and it fell away. It's hard to remember the last negative thought, hard to remember the last upheaval. It's like a dream.

CH: Is there anything that people can do—all need to do—to prepare for this shift?

An: I think the Dimensional Shift has always been there, we are just dropping away the layers that kept us separated. Once we let go of those things that don't serve us anymore, we will be there.

CH: How do these layers show up for people? What do they manifest as?

An: Even the smallest ripple, when I am in my calm self, I notice every little ripple.

CH: Is this why people are having so many emotional issues from the past coming up? Is it the clearing?

An: Yes, because we are all parts of each other, we are experiencing each other's ripples. Part of the realization is to realize that it's all good; it's all part of the connection. We just have to simplify. The more we let go, the easier it is.

CH: What if we are in a relationship and our partner is having ripples? How do we deal with that?

An: By not taking it on as your stuff, not taking it personally even if it's coming from you. Just feeling the ripple and not taking it on.

CH: So the outer reflection of this cleansing is the hurricanes and tsunamis, etc.?

An: Yes, it's just the outer reflection, and as we start taking on the ripples we just let it drop away. It's time to let go, and looking back we will see that it was something it was time to let go of.

CH: How are our relationships in the New Earth different to the way they are now?

An: They are closer and deeper and we don't react anymore. We are just joyous and loving, and we recognize that there is nothing to do except to enjoy each other.

CH: Is it more of an unconditional loving connection?

An: Yes, everyone does unconditionally love each other and they are appreciating that.

CH: What happens to the body in the New Earth when we are out of the element of time? How does that affect the body?

An: It doesn't feel like there is as much gravity. It's almost like gravity and time worked together, but without the gravity and time you feel that nothing is holding you back. You are not encumbered by anything.

CH: When it comes to the time to release the soul from the body, what does that look like?

An: It's almost as though the soul is already free and it's really joyous. There is no difficulty about coming back and leaving. There is no effort. Just ask and sing your joy with not even a thought of coming and going. We are happy with every condition, with or without the body.

CH: So after the shift do we still have an aging process in the body?

An: No, it's more like how we used to be. There was a time when we were very free coming back and forth and at some point we got caught by gravity. It was different before that; we were free.

CH: Was that what we call The Fall, coming down into density?

An: Yes, it was getting caught up in gravity.

CH: In this third-dimensional time our bodies have dysfunction, pain, bits that don't work. What happens to those bodies after the shift?

An: It's a memory, but it's very like a dream because everything feels more real without the pain. The pain felt like an experience but that's all it was, just an experience; there was no truth to it. There is memory if you choose to look at it, but there is no charge around it anymore.

CH: Looking at the timeline moving forward from now, do you see when this shift is going to happen?

An: It's happening now but I do feel that there will be a time when I just walked through that door when we are all looking at each other and noticing. We are surprised. We are trying to remember when we had our last negative thought. We think, "Well, I've had a few, but it's been quite a while since I had my last one." The negative thoughts were those ripples. We are all surprised and happy. There is a point where we all look at each other and try to remember when our last negative thought was. It was a while back.

CH: So we move through the shift as a group, a lot of people together?

An: Definitely. When it happened the shift was just there and we decided to join it.

CH: Does everyone make the shift?

An: There is no one on the other side of the door; everyone is on the good side of the door.

CH: Were there some people who chose not to make the shift who went somewhere else?

An: It doesn't feel like anyone is missing, but that's because everything is whole.

CH: So on the New Earth there is only wholeness?

An: All the energy is whole. Your home and everything else is whole too.

CH: What does this have to do with the mission of the Light-workers?

An: I know that I came to bring everybody with me.

CH: So you knew when you came from the Light that you would wait until everybody was ready to go and then you would go together?

An: Yes.

CH: Why did you come to Earth after all? What was the mission?

An: I knew that this was a unique experience. I like that feeling, moving things into a higher vibration; it feels like part of my purpose is to do that.

CH: So was lifting a physical dense body into a higher vibration the purpose of the whole journey on the Earth?

An: It feels like density was part of my body, but now I am on the other side of the door and it feels that it was appropriate to have had that experience.

CH: Was this part of your soul's evolution, to have this whole experience of density, pain, and suffering and to return to the Light?

An: I think I always knew that was the goal. There was just a period of time when I chose to have the experience of being separated. A part of me always knew that the culmination of the experience was to return to Oneness.

CH: So what is special about this lifetime?

An: It's a harder, grosser, heavier experience. I think the release from it is special too because we are going from such a slow, deep, long drawn-out experience to a very light and unencumbered experience.

CH: In a way we fell down to the deepest darkness, and we are coming back to the Light from that place, is that right?

An: Yes, looking back it seems like a dream; it was just another experience but it was unique.

CH: Is there a message that you would like to give to the people of the Earth who are feeling discouraged, overwhelmed, and afraid?

An—It's time to just give up to your Higher Self everything that doesn't serve you and let your Higher Self lift you up. It's hard to do that by yourself, but your Higher Self has always been there to help whenever you choose. Just ask and it will be.

Chapter 14

How the Shift of Earth's Axis Will Affect Those Here

D is a powerful energy worker living on Maui. She is an ancient Lemurian priestess who has chosen to come back to assist the Earth through these changes. She travels widely around the planet helping the Earth grids to restore their frequency to the Oneness we had in Lemuria. She is very much connected to the dolphin and whale families. Her journey with me began with a visit to the Lemurian crystal light city that is in La Perousse Bay on Maui where the dolphins swim.

Charmian—See yourself standing on the shore at La Perousse Bay looking out at the Crystal City that is shining in the sun on the horizon. Do you see it?

D—Yes, I see it.

CH: So imagine yourself to be floating out across the bay walking upon a beam of red-gold sunlight, which transports you right to the heart of the city and up to the top where the great temple is. You are standing underneath the crystal dome receiving the golden rays from the sun through the domed roof. The rays come down through the crystal spire that points up from the center of the dome towards the heavens. As you begin to feel that energy, tell me what is happening.

D: It feels like a dome that is capped. I see a cap on it, and I am in a bubble.

CH: Feel the golden light coming in, it is the light that is the essence of who you are. What do you see or feel as you receive the golden energy from the sun?

D: It feels very familiar; it is natural for me to be covered in golden light.

114

CH: Within this light let us invite your Higher Self or your God Presence to come into the bubble with you. She has a different feeling than the energy that is D. What does that feel like?

D: Like an overshadowing from outside the bubble looking in.

CH: Invite her to come into the bubble with you.

D: She won't fit, she is so huge.

CH: Then expand the bubble; make it bigger so she can fit in.

D: The bubble has burst. She is huge. She is bigger than the Earth.

CH: She is your God Essence, your consciousness as it came from the Source, which is the Essence of you. She is the one who we would like to guide this session. She knows your past, your present, and your future on all planes and all dimensions.

D: Her name is Aurora. She is beautiful, like a rainbow, luminescent with flecks of light.

CH: Let us ask if she will be willing to guide the rest of this session.

D: Yes, she will.

CH: So let us ask D's personality self to drop away now; the ego self is dropping away so there is only the Higher Self here.

CH: Why have you chosen to come to D at this particular time?

D: There is no time.

CH: What would you like to gift D with at this time on her earthly evolutionary journey?

D: The capability of being able to feel the Earth in the palm of her hand, as small as it is for me. It is so manageable then.

CH: Let us look at the time of December 21, 2012, the time of the Earth's alignment with the galactic center and the great central sun. What do you see?

D: That is where she comes from. She is the center. She is Source consciousness.

CH: Let us look at what happened on the Earth when this alignment took place.

D: There is so much light pouring in. It is blinding, so bright I can't see it clearly.

CH: What is the quality of this light?

D: It is deflecting meteors like dark lumps of rock from hitting the Earth. There are people who throw stones at the Earth. There are some consciousnesses that are trying to stop the Ascension of the Earth.

CH: So this shield has been put in place to prevent that?

D: Yes.

CH: Let us look at the light that is getting through to the surface of the Earth, not the shield, but another energy that comes through from the galactic center. What does that look like?

D: It's like a spiral rainbow helix with rainbow colors and flares with flashes of light coming off it. It looks like it is shifting the axis of the Earth. It is tilting it the other way. It is going a different angle.

CH: What is the effect of that?

D: It will bring climate change because instead of the sun being at a particular angle it is going to be hitting at a different angle, perhaps not directed at the equator anymore. Then one whole side, even the South Pole, is going to get a lot of sun hitting it. It hits one way and when it turns another place is more directly facing it so cold places will become warm. It is tilting on a different axis. I am seeing it quite clearly.

CH: That is what happens on the physical level, but what effect does the rainbow ray have on people who have been preparing and are ready to raise their frequency?

D: It brings people together more. I think they have wanted to come together more. Like when the tsunami and earthquake happened at Fukushima, when there are disasters, people come together more to help and to care for each other, rather than just living their separate existences and trying to make things work well for themselves.

CH: The ray also it affects people's physical bodies, their light bodies, and their DNA. What happens to D's body when the rainbow spiral helix comes down?

D: She gets younger, lighter, and happier. There is someone who is going to really love to be with her. I see a very beautiful, peaceful, masculine man.

CH: Is this after 2012?

D: After she gets really light, around March 4, 2013.

CH: What about the other people who receive this activation? How does it change the way people live on the Earth?

D: They will be able to feel the fiber optics between them, and they will be much more palpable, like lines of light connecting everyone together. They will be able to find each other more quickly and easily. The network between people will be much better established. People will be much more willing to connect telepathically and through the Internet and Skype. Everybody will be getting much more socially networked through the media, and it will become exponentially easier for people to find each other. Like dolphins, we will trust our radar and sonar more. Dol-friendly people will find each other no matter where they are on the planet.

CH: How will people treat each other? How will their consciousness change?

D: It will be more like a pod mentality, where a pod plays together, feeds each other, and lives together in a lovely snuggly, warm consciousness. This type of dol-friendly behavior is coming back on the planet. There will be more energy connection, eye contact, and being able to see deeply into each other will be much more socially acceptable.

CH: How will people relate to the Earth and the environment?

D: We are sacks of water, mostly space and water; we will be releasing the body from being perceived as solid. There will be a shift similar to when people thought the Earth was flat, then discovered it was round. It is a new paradigm where we know that we are a watery skin bag and we are luminous, light, and mostly space. There will be much greater understanding of ourselves as energetic frequencies rather than solidity, scarcity, and security.

CH: How will people sustain these younger, lighter bodies? Do they eat in a different way to the way they eat now?

D: Most of what people now consider food has no nutritional value, no light, and no energy. It is actually dulling and deadening. Of course people will need to eat less, and what they do eat will be more for sustenance than comfort. They will intuitively eat according to what is elevating to their frequencies and is recharging them. They will eat what they need to recharge rather than to numb or dull themselves out. They might eat a little for taste but mostly to build their molecular charge, more green food and alkalinized water. It's all about recharging. Like a battery we need pure water and we need lightening up the density with lots of stretches, lots of yoga and aerobics.

CH: Is there anything different about the way food is produced compared to how we do it now?

D: We do it with clean air, aeroponically. It needs lots of nutrients and less soil where it is produced and pure, clean water rather than the radioactive air and water that we have now.

CH: How do the new children fit into the New Earth that is rising?

D: They are our children or our children's children so they don't have the samskaras (past life impressions and experiences to be cleared) that we have. They don't have the karmic completions that we have had. They come in ready to "beam" unless we destroy them or ruin them. They are ready to "beam" on this level because they don't have limitation or fear in the way that we had it in the past. People do not suppress them so they don't have to turn to drugs or alcohol. They have already been exposed to mind-altering states through spiritual practices or ceremonial practices. They are already born to shamans or to sons of shamans. They are already magical shamanic kind of beings themselves. They are just with us and there is no separation. Twenty-seven year old boys don't really see an age difference with us. Some of them are already so brilliantly skillful that they can make corporations or design new things or come up with ways to organize things that will be very prosperous.

CH: What will technology be like in the New World?

D: Much more wind and solar energy. There will be hydraulic energy from water and solar.

CH: Like hydroelectric power?

D: Yes, or taking seawater and de-salinating it, making it fresh water. Everything is renewable and recyclable, and the Earth itself has frequencies that Tessler knew about. There are clean ways of using frequencies to create power. It's not good to dig into the earth and disturb her molecular structure; this makes her angry and creates disturbances such as volcanic eruptions, earthquakes, and tsunamis.

CH: Let us look at the effect of the Earth tilting and the interruption of ocean currents. What happens to the people who have not accepted their rainbow helix? What does life look like for them?

D: They receive the energy, but they just don't know how to integrate it or deal with it.

CH: What happens to them?

D: They just disappear or they die; they go crazy. They are resisting it somehow. It's really like the pink elephant in the room. It's there but they just don't see it; it's like having candy there but you have to suck on it. You have to plug into it, the energy is there, the sun is there but they are not plugging into it. They are trying to operate in the old way. Other people operate bringing in the new way. It drags you down to operate in the old way because it feels like you know the new way and the old way makes you feel sleepy, it drags you down.

CH: So there is a difference in the vibrational rate between people in the new way and people in the old way?

D: It is hard for people in the new frequency to get pulled back into the old way. If they stay in the positive frequency with other like-minded people, they could have

fear that they might do something from the old frequency.

CH: Does that mean that for a time the Earth will have two
frequencies?

D: Yes, like Moses and the Promised Land, the old people
who want to operate in the old way are going to die off
and the young people come together. The world is such
a melting pot, and there is so much technology. Everybody will Skype each other and see each other as a lovely
interesting human being. They will see the world as one.
They will see that we that we live on a lovely blue planet
that we have to take care of. There will be no more judgment, and all that kind of stuff that keeps us separate
from each other will disappear. People will go to China
and see that they are really nice people. There are people
in North Korea or certain Arab or Muslim countries that
are being indoctrinated and are being very fanatical.
Amongst their peers there could be groups of fanatics
like bandits. They have guns like in the Bad Max movies;
there could be terrorists and robbers on the road. That
could happen.

CH: After a while when the Earth has stabilized at the new
frequency, what happens to the energy of conflict, war,
and division?

D: That is a long time from now, but eventually everybody
wakes up. It is hopeful that everybody will see the Truth
that we are all One.

CH: Isn't that what the shift is all about when the rainbow
spiral helix comes to the Earth?

D: It's not overnight; it takes two or three hundred years,
but it is moving in that direction.

CH: Will some people will go there immediately if they have
prepared?

D: Yes, but not the whole planet.

CH: What will be the role of those who have prepared and
who have received the new energy after December 2012
when the Dimensional Shift has happened?

D: To keep on coming together in more organized and fun
ways; being more willing to pull together in tribal ways.

CH: What would be most beneficial for D to do between now and the time of the shift? Are there places for her to visit to receive new frequencies?

D: Yes, Machu Pichu and Vale Bamba; maybe Southeast Asia, Thailand, Egypt—but not this year.

CH: Is there anything she needs to do for her body to prepare physically?

D: She carries the weight of the world on her shoulders but as others take responsibility it will fall away from her.

CH: Do you have any last message for D for assistance and guidance between now and then?

D: She is really loved by the Masters and Source. She is doing high-powered, high-level work on this planet. She can't even believe it or know it's true. This is good today.

D: The energy is knocking me out it's so powerful.

CH: Just receive the energy and let it integrate at all levels of your Being.

Chapter 15

Return of the Christ

*A*ngelina is a very light Being. She embodies the Divine Feminine and holds events where she brings in the frequency of the Divine Mother. People sit and receive the blessing of the Presence. She is an old Lemurian soul who has come to assist with this transformation of the Earth. I guided her in a hypnosis session through the Dimensional Shift into the frequency of the New Earth.

Charmian—We know the energy of the Earth is moving into a new frequency. Let us track forward to the time after the Dimensional Shift. Where is Angelina after the shift?

Angelina—It will be warm, a warm ocean.

CH: How is the Earth after the Dimensional Shift? What does that look like?

A: Light.

CH: Focus on the light. Look at your own body. Tell me what that looks like?

A: It is just energy—warm, soft, light, and beautiful.

CH: How is it different from the body you are in now?

A: It is much more peaceful. It's so light; it is warmth in itself.

CH: What does Angelina do with her Light in the New Earth?

A: She is sharing it with others and assisting those who need a bit more time.

CH: Look down at the Earth as if you were flying. How does that look?

A: I am in the formlessness.

CH: How are people living now that is different from the way they were living before the Dimensional Shift?

A: More in community. They are all together; they are taking care of each other. There is Love, and everyone seems to be more taken care of. They walk arm-in-arm smiling, and they are doing wonderful little tasks. Everyone is more at ease.

CH: How do they create the things that they need?

A: Everything is given to them with their own energy. Whatever they need is there through Grace. It's not even asking; just an inner focus on what is needed right now and it is there.

CH: Are there any machines or cars?

A: I don't see that.

CH: How do they travel?

A: They walk, they ride horses, and there are Light ways of traveling. They can travel with their bodies to other places. It's effortless to get somewhere.

CH: Are people still in their physical bodies? How are they different from the way they are now?

A: They are so light, and they can just move by energy. It's like flying—they move from one place to another. I also see white horses.

CH: What is the relationship now between animals and humans?

A: It Is One. I can see this beautiful white horse, and there is no fear between us.

CH: Is there telepathic communication?

A: It doesn't need words; it's a very deep understanding of each other.

CH: How do we sustain our bodies?

A: Like Adam and Eve, very simply. We eat fruit from a tree, and we are much more Light. We are going to be eating really simple food that will grow around us. We don't need to eat much. No more proteins, just God's food, herbs and apples from the trees.

CH: Look at the trees. How are they different, and how is our relationship with them different to what it is now?

A: They produce beautiful fruit, they are lighter, and there is so much abundance and beauty.

CH: Can you see the Tree Spirits?

A: I can see the Oneness with the tree, the Love.

CH: Is that why everyone can live together in harmony and love, because they are in Oneness consciousness?

A: Yes.

CH: How did we go from where we are now with war, financial crisis, and pollution on the Earth to this Garden of Eden?

A: Through the Ascended Masters and everything in the Light world. Mother Mary has a very big role. She is destroying a lot of the darkness, and the Divine Mother is helping. Humanity has suffered in the past. We have gone through this cycle, and they are assisting us in this process. It cannot be without pain because we all have to detach ourselves from form to find happiness within the very depths of the heart. In all of this we have so much help that is available. We just have to ask for it.

CH: When we go through the shift, are we still in physical bodies?

A: We are in a physical Light Body, but we are still here on this Earth. We are in a physical body, but we no longer feel the density of it.

CH: What happens to the physical vehicle when the Light Body comes in?

A: It heals; a healing takes place in the organs. We still have a body; it's just not so dense.

CH: Does the body still age?

A: There is no aging. It's like this: imagine an old lady who has lived many years; if you looked at her now she would look younger than she did many, many years before because the Light is shining through so it doesn't matter about the body. It is ageless.

CH: What is the Light that is shining through the New Body?

A: It is the Christed Self.

CH: Let us look at the moment this Christening happened on the Earth. Did something happen to trigger it?

A: There are things that are happening to trigger this Light. Many things are happening on this Earth that will bring us as Light Beings into ourselves. It comes all of a sudden by Grace, even in the form world where everything gets very shaky.

CH: So we go through some kind of Tribulation?

A: Yes we do and we have to start now because we are feeling it already. We have to tune into the help that we can get from the other side. We need to begin now to practice staying in the Presence, to be prepared.

CH: What form is the Tribulation going to take?

A: Pain in the world. Lots of suffering, people crying, and losing loved ones. It's going to be different everywhere—different things—but we don't have to worry because we will be safe. We will be in our Higher Consciousness—it's different. We feel different. There will be earthquakes and tsunamis, but we will rise above them.

CH: Will there be a moment in the Tribulation when Grace comes in and lifts us up?

A: We are there to assist others and that is how we get the Grace because we have compassion to assist those who are in fear.

CH: Do you see many people leaving the body, leaving the planet?

A: Some will leave and some will stay.

CH: What do we need to do right now to prepare for the Tribulation and the Dimensional Shift?

A: Try to live out of our Essence as much as we can so we can live in purity and Love. Try not to get so caught up in all the stories and behavior patterns that are going on out of fear right now. Tune into the Divine Mother—how does she want us to live our lives from moment to moment? Being with each other, we need each other right now more than ever before. It's not a time to be a recluse. It's a time to share.

CH: Can we share this information to prepare our children?

A: They are young. Just hold them in your hearts and let them know lots of things are happening but everything will be fine and not to be afraid. Give them mothering and nurturing.

CH: Do you see the timing when the Grace comes in?

A: It's getting very close. We are making it happen now.

CH: Do you have a message for those of humanity who are awakening right now?

A: Don't worry, be happy, just trust, trust and say, "Yes." Release everything and let it go. It is the old that is going. There are many in our lives who are helping us to dissolve our old patterns. Give it all to the Light.

Chapter 16

Earth Changes Are Preparing the Way

George is a physician living in Hawaii. He is an Atlantean crystal healer who is beginning to rediscover his gifts. He saw a tidal wave hitting the coast of Los Angeles at the time of the alignment with the galactic center. What I have learned from other sessions and from my channelings from Mother Mary is that the severity of the geological events that are part of the energetic shift is dependent on us. If humanity can align with the necessary changes that will bring us out of conflict and into harmony then the catastrophic events such as tsunamis and earthquakes do not need to happen. It is up to us to work consciously with Mother Earth to assist her to raise her frequency through our prayer and meditations. Then she won't need to shake so hard to bring about her birthing into light. All of the large events such as the earthquake in Japan, the hurricane in New Orleans, and the giant tsunami in Indonesia are for the purpose of opening the global heart chakra, reminding us that love, community, and family are what are important, not material things.

Charmian—Let us go forward to the time of the Dimensional Shift. Can you see when that is?

George—It begins in 2012.

CH: How is the Earth different from our time now?

G: People are more spiritual. People are more peaceful and loving. There is no need to eat.

CH: How do we sustain our bodies?

G: We sustain our bodies from source energy. We breathe prana into the heart from Source.

CH: How is the air quality in that time?

127

G: Everything is clean, there is no pollution, and there is no dust in the atmosphere because plants have grown over all the places where the dust came from.

CH: What happened to the factories?

G: They just stopped because we didn't need them anymore.

CH: How do people feel in the New Earth?

G: They feel full of joy and light.

CH: What happened to the pain, the anger, and the grief in the old world before the shift?

G: They all disappeared at the shift.

CH: Let us look at anything we may have done to help to create this shift. I am seeing you and I in the Lemurian crystal city off the coast of Maui. We are in a control room with lots of panels with lights, buttons, and screens. We are both there in our Lemurian light bodies. You are turning dials, programming events that need to happen before we Ascend. There is a dial with dates like a calendar. You turn the dial to the earth after December 21, 2012. The planet appears on a big screen. There is an alignment with the galactic center that sends a beam down to the planet. What do you see?

G: It seems more pleasant, more loving.

CH: Look at the surface of the planet. How is it different?

G: It looks like there is a tidal wave. There is a big wave heading for Los Angeles.

CH: What happens when it hits?

G: It causes a lot of destruction.

CH: How far does it go?

G: It goes through most of the city.

CH: What happens when the water recedes?

G: Everyone has to clean up the mess. It is a cleansing.

CH: What was the purpose of the tidal wave?

G: It was a cleansing to get back to the premise of love.

CH: How did the tsunami help with the return to love?

G: That's what happens to those who are not in harmony and love.

CH: How do people live in that area after the tsunami?

G: It's more about all the Earth changes not just the tsunami. People realize they need to live in love and harmony.

CH: Let's take a look at Los Angeles one year later. What does that look like?

G: It is back to the ways of Lemuria, people are able to manifest what they need.

(CH:I have been shown in other sessions that these dramatic cataclysmic events can be averted or diminished if we work consciously to assist Mother Earth to cleanse herself through our prayers and our love for her.)

CH: I am seeing community gardens. People are living in communities around the gardens instead of all spread out in lines and streets where you don't have any connection to your neighbors or any sense of community. People are all growing food together and having a lot more interaction with each other because they are not going to work every day like they are now. They are spending more time meditating, praying, and connecting with animals and plants. They are cooperating with the garden spirits and the nature spirits to produce food. They are singing and dancing and celebrating a lot more—celebrating being alive.

CH: Do you see what is happening in the rest of the world?

G: It seems to be all the same frequency.

CH: How do people treat each other?

G: Well, that's part of it. People treat each other with respect and love.

CH: Do you have a special area of focus and expertise in the new world?

G: It is with crystals.

CH: What do you do with them?

G: I work with the crystals. I am directing their energy.

CH: Do you work with people or the land?

G: I work with people. I am doing healing with them.

CH: How soon after the galactic alignment does this happen?

G: Straight away.

CH: How do you know how to do it?

G: I just know, it's just there.

CH: How is it activated?

G: It's just available.

CH: How do you feel when you do this work?

G: It's a good feeling and it's a contribution.

CH: How do you feel after the shift that is different to the way you feel now?

G: I am more peaceful.

CH: Are you happier?

G: Yes, because I am peaceful. I have an inner peace.

CH: Ask the Self who has already made the shift what George can do between now and then to prepare for the energy shift.

G: Meditation with the crystals I already have.

Chapter 17

Peace Shall Reign

*A*lani Purplebird is a very gifted artist who lives in Hawaii. She has a strong connection to the fairy and the elven realms. She creates beautiful fairy costumes and accessories that help people to connect to the magical kingdoms.

Charmian—Let's go to the time after the Dimensional Shift. Where are you?

Alani—In a beautiful home overlooking the ocean full of light with lattice windows.

CH: What has happened to the Earth?

Al: People are walking on the beach with beautiful colored sarongs. They are happy. They can all come out now, it's free to come out.

CH: Is it safe?

Al: Yes.

CH: What happened on Earth to bring about this change?

Al: There was a global catastrophe.

CH: Did it lead people to the way of peace?

Al: The people on the Earth now are the ones who have been waiting for love and peace on the Earth, those kinds of people.

CH: What are you doing on the New Earth?

Al: I'm painting. I have friends around me.

CH: How is your body different to the way it was before the shift?

Al: It is slimmer. I feel at ease because I am wearing a sarong and it is very comfortable.

CH: Does your body feel lighter?

Al: Yes, it feels relaxed.

CH: Do you need to eat food in the same way you did before the Dimensional Shift?

Al: We eat lots of fruit. I see a big bowl of fruit in the house.

CH: Do people eat animals?

Al: No.

CH: What happened to the physical problems Al had before the shift happened?

Al: They released, it was a release.

CH: What about your emotions? How do you feel?

Al: Like a storm has passed.

CH: It is a relief?

Al: Yes.

CH: What does Al need to do now to prepare for the energy shift?

Al: Eat salad, not too much fruit, it is too acidic.

CH: Do you mean cut down on fruit or balance it with alkaline foods?

Al: Eat more alkaline foods for balance.

CH: Let us look at the timeline. If we track from where we are now to the time of Ascension what was Al doing to prepare for this event?

Al: Blessing the Earth, praying every day to Divine Mother, staying connected with the Divine Feminine.

Chapter 18

Eye Hath Not Seen

"No eye has seen, no ear has heard, and no mind has imagined what God has prepared for those who love him."

1 Corinthians 2:9

R is one of the original Lemurian grid masters who brought the frequency of the Oneness to the Earth. In her session we went through the gateway into the New Earth and saw the changes that had completely shifted the awareness of humanity. Her Higher Self Isabella came in to connect with her and then an even higher aspect of herself as pure light beyond form came and agreed to answer the questions I put to her.

Charmian—See yourself going through the Ascension process. Tell me what that looks like.

R—It's like a death but not a physical death. It's a letting go, a loss of old beliefs. It's like losing a shield. A heavy, dense shield of whatever no longer serves me, it's falling away. The density of old beliefs, old energies, fear, it's giving way to the Light. I see many strands of color; it's like the DNA strands, it is running light, strands of light, energy, and quickening. Sound; I no longer have a body. I am the strands, I am the Light, I am this amazing energy. Even this is temporary, only a transformation, a temporary pathway.

I Am All that I Am. I Am pure Light. I Am All That I Am. I Am One. There is no division, no separation. I Am That. I recognize Who I Am. There are many, many, many that are awakening to the Oneness. We recognize that we are all One. We are interwoven. We are One with our power, we know Who We Are. We are God. We have chosen in this lifetime to embody this. We will remain in form, we will take this Oneness and through this consciousness we will create euphoria. We are not limited by form; any form does not limit us. This is what we shall be sharing with others. We are limitless Beings. This is what God has always wanted us to know. As we become God, we know this. We are unlimited so we are free to create. We will create from a place, a consciousness of pure love and pure light, pure bliss.

CH: How does this relate to the alignment of the Earth with the galactic center on December 21, 2012? Does this date have significance in this process of returning to consciousness of the Oneness?

R: For us there is no time or space, for us this date is not significant for it is already occurring.

CH: What about for humanity in general? Is this date significant?

R: There is a network of Lightworkers. They are working on many, many levels that you know as the light grid. Some are the Pleiadeans. There are so many forms. So many have made the agreement to focus on that day you call December 21, 2012. From the collectiveness, from the All That Is and from that focus there will be a major shift that will affect the planet Earth. There will be many that will awaken and many that will leave.

CH: What will our role, those who are already awakening, be at that time?

R: We ask you to focus on love, we ask you to focus on the light that You Are, to know yourself and claim yourself as God. This is important to anchor this vibration. It is important to release any seeming fear that may arise from the changes that may occur. Be One with your knowingness, hold the Light, Be One, be the anchor. On this day there will be much energy and light that will be concentrated. It is important to breathe and anchor it into the Earth. It is important to gather together with like-minded ones so that you can serve as the anchors of the energy that is being transmitted by many. Be light, light-minded. Stay centered in this Divine Love and know that there are many, many beyond your imagination, many Lightworkers who are involved and All is Well.

CH: In view of the immensity of the changes that are coming, where do we need to put our financial resources?

R: We want to speak to you about the predictions you have heard. We cannot say if they are true or not true, for nothing is set in stone. For us there is only one truth and that truth is your own awakening to pure love, knowing yourself as pure light, knowing yourself as God. So we say that your truth is your truth and we cannot predict what that may be. We also recognize that there is race consciousness, if you will. We know that there is only one consciousness and that consciousness is God.

What we are trying to say to you is, "It's OK to be prepared." We feel that All Is Well, we feel that really no preparations are necessary; you will be guided in the absolute moment, at the right time if there is something necessary for your existence on Earth. We feel that humanity and each one of you needs to be in trust, to know that you are One with God. In this moment you are always provided for, watched after, for you are a part of God and God is a part of you. There is nothing less, nothing of your wanting, everything that God is You Are, God lacks no-thing, my dear one. So you see there is nothing that you lack for you are God. You lack no-thing. If there is a need you will be notified, you will know. I promise you this. God loves you and there is nothing that God holds back from you.

CH: Would it be helpful to put our financial resources into land and into living in community with like-minded people on the land?

R: Yes, we would absolutely love to see you living in community. We think this serves you well. Everything will be shown, everything will be given to you, and it doesn't need to be figured out. It will unfold in its natural unfoldment. Everyone contributes from his or her wholeness. It will unfold in its beauty. It is beyond your wildest imagination what is going to unfold all through you. It needs all of you to make it but it will be a work of art and each individual—I don't want to say individual but as you see it on your Earth as individual—it takes each individual to contribute to the wholeness of your community.

We do say that there will be some earthly changes. You will see yourselves pooling your resources and your talents and you will understand interdependency. At this moment you know only to be independent. This is not Who You Are. These changes are for you to learn that lesson, interdependence. Each is important to the whole. Each has been given a gift. Each is a Divine expression of God, and you will know this to be true.

CH: Is there anything you can share with others, with our children, with our families, with humanity that will help them to prepare?

R: Do you remember when I said that truth is truth? You see truth is truth to the one it comes through. Their truth is their truth, and so their path will be their path. So what we would say to you is to have them meditate, to do all that they can to remember Who They Are. This is very important so that they don't go down a path of fear. It's good to come together in groups of like-minded people to help the vibration. It is very important at this time for people to pay attention to their intuition and to follow it regardless of doubt; do it anyway. We also ask that you pay attention to synchronicity especially in threes. Water is important; we do suggest a cache of water.

CH: Is there anything that will help R to prepare for the shift?

R: Lay crystals on your body, be in salt water on a regular basis, and work with crystals. This will help to cleanse the denser vibrations and help you to be more receptive to the energy and the vibrations that are coming in. It is important for you to be in a place of joy, anticipation, and excitement for your prayers are being answered. You will experience heaven on Earth in this lifetime. You will experience a Divine love that is beyond anything you have ever conceived, know this to be true. You will be in such great joy and bliss; you will have such telepathic experiences with animals, people, dolphins, and whales. You will see colors and vibrations and things will be instantaneous. You will experience yourselves as unlimited Beings, my dear ones, and you will be in great joy. So rejoice, rejoice for you will be unlimited. Breathe that in and know that I speak the truth. In your lifetime it will happen, and you will be in partnership experiencing this.

Conclusion

We can celebrate now. We have come through. We have brought the planet through the darkness, and the dawn is here. We are the children of the light; we are infinite power and infinite love. We have prevailed. Despite all attempts by forces who have tried to hold the Earth and all of humanity in darkness and suffering, we have come now to a point of no return where the Earth is shifting into her new frequency of Oneness and unconditional love. We are returning to Grace and to the Garden of Eden. The prophecies are fulfilled. This is the end of time as we know it, and now we are lifting the Earth to the frequency that we know as Home. This is the time when we come into our own. We have maintained the integrity of our Beingness despite all attempts to disconnect us from our Source and to remove our influence from this Earth.

This event on the Earth, her Ascension into her Light body, is the completion and fulfillment of our mission here for this whole cycle. None of us need to reincarnate here again unless we choose to. We have been released from the wheel of death and rebirth, from all karma, and we are free. We can choose to stay on the Earth to enjoy the New Age of Gold that we have helped to build, or we can go Home to ourselves as light essence.

We are in the last days and there is still some cleansing to be done, but the outcome is already determined and there is great rejoicing in the heavens. Each one of us who chose to come to Earth to create this shift has our name written in letters of gold in the book of records, the book of love, and we can be proud. Against all odds we have prevailed. So be kind to yourselves now as your physical bodies are working so hard to prepare themselves to become bodies of light. We need to be gentle with ourselves as our emotional bodies strive to release themselves from eons of pain and suffering. We have spent so many lifetimes on our knees before the altars of Gods and Goddesses, berating ourselves as unworthy. Now we are awakening to the knowledge that we are the Deities that we have been worshiping for all of these years.

We are the ones on the altar to whom we have been offering our prayers and devotion. We are the Creator Gods and Goddesses who have lifted this planet out of darkness and into light.

We are almost there. As the electromagnetic frequencies continue to increase on the planet, we will reach a jumping off point where we shall be lifted up into our own Christed being "in the twinkling of an eye." Just one small step and we will have passed through the doorway into light with no turning back.

I am humbled and honored to be walking this path with you. I celebrate each and every one of you. We shall sing and dance together in the garden of love that is the New Earth. She has risen like the phoenix from the ashes of the old. Reach out and ask for what you need. There are legions of angels waiting to celebrate you and to serve you. You are the Beloveds; you have come Home.

December 21, 2012, was the trigger, and since that time we have passed through many portals that have increased the light frequency of the Earth and all of her inhabitants. As the light increases, it brings to the surface all that is not of the light. As we look around the planet, we see so much violence and conflict, and within our own environment there is much to challenge us as outmoded relationships or ways of being that are not in alignment with our soul purpose here that need to be released.

Remember that you are in charge here. Your own soul is directing your script, and it is all bringing both you and the planet back to love. Out of chaos comes harmony, out of conflict comes peace, and out of hatred comes love. Each equinox, solstice, and eclipse is bringing us one step closer to the New Earth, but when we get there is up to us. When enough people are ready to hold the frequency, then the whole planet will shift. So raising your vibration is the most important contribution you can make at this time to the Earth, to your children, and to yourself.

Revelations 21:1: "I saw a New Heaven and a New Earth, for the Old Heaven and the Old Earth had passed away."

Contributors

Charmian Amarea Kumara Redwood

Charmian is available by phone or in person for hypnotherapy sessions to connect with the soul or Higher Self. The soul is able to give information about all issues and contracts from the past, present, and future lives of the client. In the session multidimensional aspects of yourself from past or future lives or other star systems and planes are brought forward to be integrated into your current life. She offers a guided experience of going forward into the New Earth to experience what it means for you. She has created many guided meditation CDs and her book *Coming Home to Lemuria*, which are available through her website.

www.cominghometolemuria.com

Keala Gerhard

Keala is a very powerful healer; she uses Reconnective Healing to align her clients with the new frequencies of the Earth. This technique helps the client to be lifted up into his or her light body.

Reconnective Healing Practitioner. Teacher for Hawaiian Spirituality

http://www.kealahealing.com

Britta Lehnert

Britta is a very talented artist who paints and creates beautiful sacred jewelry with crystals and semi-precious stones.

Simple sophistication to boho chic jewelry and beads.

www.lovebritta.com

Kathy Williams

Kathy is a certified medical intuitive and yoga therapist who helps people find optimal health through breath, movement, and meditation. She is also a practitioner of I'O Ancient Hawaiian Energy Healing.

Contact Kathy at coachkathy@gmail.com or www.yoga4backpain.com.

Colin Whitby

Colin runs the popular webzine, "The Magic of Being," and has his own website. As an author he has contributed to *2012: Creating Your Own Shift* and *The Sacred Shift: Co-Creating Your Future in the New Renaissance.* The last year has seen some massive shifts in energy for us all, and many more shifts are on the way as we move into an ever lighter and finer vibration. As an energy route finder or mapmaker in this new light, Colin likes to help us navigate in this often strange and wonderful ocean of awareness.

www.MultidimensionalAlchemy.com

Susan Michelle Moll

Susan is a spiritual intuitive and clear channel for Divine energy transmissions. She is author of two spiritual books, *Wege der Bewuesstwerdung,* published in Germany, and *The Magical Princess,* a spiritual book for juveniles to be published shortly by Tate Publishing & Enterprises LLC.

www.theislandoflight.com

Angelina Vey

Angelina is a channel for the Divine Mother to ground on the Earth. She gives satsangs where she transmits the frequency of Divine Mother to all who are present.

www.angelinaoflight@yahoo.com

Beloved Love

Beloved is a very talented singer, songwriter. She connects to the inner planes and creates enchanting songs that help to awaken her listeners to their Divine Presence.

www.purebelovedlove.com

Alani Purplebird

Alani is a very gifted artist who lives in Hawaii. She has a strong connection to the fairy and the elven realms. She creates beautiful fairy costumes and accessories that help people to connect to the magical kingdoms.

http://www.fairywear.com

Mary Carol Breckenridge

As a writing consultant Mary Carol works with authors, inspiring and guiding them with the use of her technical, creative, and insightful gifts. Her copying and editing skills are finely tuned and highly appreciated.

marycarolbreck@gmail.com

About the Author

CHARMIAN AMAREA KUMARA REDWOOD

In 1980 Charmian Redwood had a Near Death Experience in which she returned to Oneness and remembered Who She Is. Since that time she has been assisting others through her workshops and personal sessions to reconnect with and their own Divine Self and to activate the DNA codes for ascension. She has remembered her many lives as a teacher and healer in the mystery schools always guiding her students to find the God Within and to empower themselves.

Since moving to Hawaii in 2006 Charmian Redwood has brought forward many memories of Ancient Lemuria where we lived in Oneness and used our intention and connection to the Source to create everything we needed in our lives. Her work now is to bring back the teams who worked in the Ancient Crystal Cities so that we can begin to create our New World. In 2012 she was guided to move to the Midwest to bring the lightcodes of Lemuria where they are most needed. She now lives in Cincinnati, Ohio.

Books by Charmian Redwood

A New Earth Rising
Published by: Ozark Mountain Publishing

Coming Home to Lemuria
Published by: Ozark Mountain Publishing

For more information about any of the above titles, soon to be released titles,
or other items in our catalog, write, phone or visit our website:
Ozark Mountain Publishing, LLC
PO Box 754, Huntsville, AR 72740
479-738-2348/800-935-0045
www.ozarkmt.com

If you liked this book, you might also like:

Dancing Forever with Spirit
by Garnet Schulhauser
The Three Waves of Volunteers and the New Earth
by Dolores Cannon
Raising Our Vibrations
by Sherri Cortland
Feng Shui From the Inside, Out
by Victoria Pendragon
The Convoluted Universe, Book 1-4
by Dolores Cannon
Let's Get Natural with Herbs
by Debra Rayburn
Out of the Archives – Earth Changes
by Aron Abrahamsen

For more information about any of the above titles, soon to be released titles,
or other items in our catalog, write, phone or visit our website:
Ozark Mountain Publishing, LLC
PO Box 754, Huntsville, AR 72740
479-738-2348
www.ozarkmt.com

Other Books By Ozark Mountain Publishing, Inc.

Dolores Cannon
A Soul Remembers Hiroshima
Between Death and Life
Conversations with Nostradamus,
Volume I, II, III
The Convoluted Universe -Book One,
Two, Three, Four
The Custodians
Five Lives Remembered
Jesus and the Essenes
Keepers of the Garden
Legacy from the Stars
The Legend of Starcrash
The Search for Hidden Sacred Knowledge
They Walked with Jesus
The Three Waves of Volunteers and the
New Earth
Aron Abrahamsen
Holiday in Heaven
Out of the Archives – Earth Changes
Justine Alessi & M. E. McMillan
Rebirth of the Oracle
Kathryn/Patrick Andries
Naked In Public
Kathryn Andries
The Big Desire
Dream Doctor
Soul Choices: Six Paths to Find Your Life
Purpose
Soul Choices: Six Paths to Fulfilling
Relationships
Tom Arbino
You Were Destined to be Together
Rev. Keith Bender
The Despiritualized Church
O.T. Bonnett, M.D./Greg Satre
Reincarnation: The View from Eternity
What I Learned After Medical School
Why Healing Happens
Julia Cannon
Soul Speak – The Language of Your Body
Ronald Chapman
Seeing True
Albert Cheung
The Emperor's Stargate
Jack Churchward
Lifting the Veil on the Lost Continent of Mu
The Stone Tablets of Mu
Sherri Cortland
Guide Group Fridays
Raising Our Vibrations for the New Age
Spiritual Tool Box
Windows of Opportunity
Cinnamon Crow
Chakra Zodiac Healing Oracle
Teen Oracle
Michael Dennis
Morning Coffee with God

God's Many Mansions
Claire Doyle Beland
Luck Doesn't Happen by Chance
Jodi Felice
The Enchanted Garden
Max Flindt/Otto Binder
Mankind: Children of the Stars
Arun & Sunanda Gandhi
The Forgotten Woman
Maiya & Geoff Gray-Cobb
Angels -The Guardians of Your Destiny
Seeds of the Soul
Julia Hanson
Awakening To Your Creation
Donald L. Hicks
The Divinity Factor
Anita Holmes
Twidders
Antoinette Lee Howard
Journey Through Fear
Vara Humphreys
The Science of Knowledge
Victoria Hunt
Kiss the Wind
James H. Kent
Past Life Memories As A Confederate
Soldier
Mandeep Khera
Why?
Dorothy Leon
Is Jehovah An E.T
Mary Letorney
Discover The Universe Within You
Sture Lönnerstrand
I Have Lived Before
Irene Lucas
Thirty Miracles in Thirty Days
Susan Mack & Natalia Krawetz
My Teachers Wear Fur Coats
Patrick McNamara
Beauty and the Priest
Maureen McGill & Nola Davis
Live From the Other Side
Henry Michaelson
And Jesus Said – A Conversation
Dennis Milner
Kosmos
Guy Needler
Avoiding Karma
Beyond the Source – Book 1, Book 2
The History of God
The Origin Speaks
James Nussbaumer
The Master of Everything
Sherry O'Brian
Peaks and Valleys
Riet Okken
The Liberating Power of Emotions

Other Books By Ozark Mountain Publishing, Inc.

John Panella
The Gnostic Papers
Victor Parachin
Sit a Bit
Nikki Pattillo
A Spiritual Evolution
Children of the Stars
Rev. Grant H. Pealer
A Funny Thing Happened on the
 Way to Heaven
Worlds Beyond Death
Karen Peebles
The Other Side of Suicide
Victoria Pendragon
Feng Shui from the Inside, Out
Sleep Magic
Walter Pullen
Evolution of the Spirit
Christine Ramos, RN
A Journey Into Being
Debra Rayburn
Let's Get Natural With Herbs
Charmian Redwood
A New Earth Rising
Coming Home to Lemuria
David Rivinus
Always Dreaming
Briceida Ryan
The Ultimate Dictionary of Dream
 Language
M. Don Schorn
Elder Gods of Antiquity
Legacy of the Elder Gods

Gardens of the Elder Gods
Reincarnation...Stepping Stones of Life
Garnet Schulhauser
Dancing Forever with Spirit
Dancing on a Stamp
Annie Stillwater Gray
Education of a Guardian Angel
Blair Styra
Don't Change the Channel
Natalie Sudman
Application of Impossible Things
Dee Wallace/Jarrad Hewett
The Big E
Dee Wallace
Conscious Creation
James Wawro
Ask Your Inner Voice
Janie Wells
Payment for Passage
Dennis Wheatley/ Maria Wheatley
The Essential Dowsing Guide
Jacquelyn Wiersma
The Zodiac Recipe
Sherry Wilde
The Forgotten Promise
Stuart Wilson & Joanna Prentis
Atlantis and the New Consciousness
Beyond Limitations
The Essenes -Children of the Light
The Magdalene Version
Power of the Magdalene
Robert Winterhalter
The Healing Christ

For more information about any of the above titles, soon to be released titles,
or other items in our catalog, write, phone or visit our website:
PO Box 754, Huntsville, AR 72740
479-738-2348/800-935-0045
www.ozarkmt.com